NUMBERS
IN MINUTES

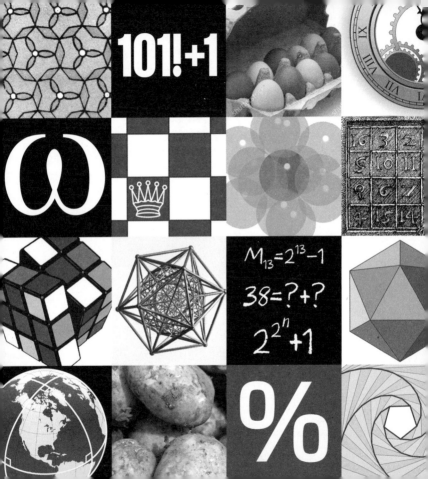

101!+1

$M_{13}=2^{13}-1$

$38=?+?$

$2^{2^n}+1$

%

NUMBERS
IN MINUTES

JULIA COLLINS

Quercus

Contents

Introduction

Numbers are arguably humanity's greatest, and oldest, invention. People were using symbols for numbers before they developed alphabets. The number twenty-nine has the distinction of being the earliest recorded number, appearing on the 43,000-year-old Lebombo bone, from the Lebombo Mountains in southern Africa, as a series of tally marks, perhaps counting the phases of the moon. By 4000 BCE, the ancient people of present-day Iran were inventing symbols for different numbers to facilitate growing trade in the region. Today we could not imagine a world without numbers – not only for counting and commerce, but as the foundation of computers and technology, and the language of science.

Mathematics, of course, is the discipline most closely associated with numbers. While most people see numbers as a useful tool, mathematicians see the patterns and beauty within the numbers themselves. Just as we may have friends who are perfect, narcissistic, vampiric, lucky, rational and imaginary, so, too, do mathematicians describe numbers this way.

This book tells the stories of 200 interesting numbers, from the purely abstract to the downright useful. It is split into three sections, each of which presents the numbers in order of magnitude. The first contains whole numbers, also known as natural numbers or counting numbers. Large numbers are given in exponential notation. For example, a number written as 3.2×10^9 means 3.2 with the decimal point moved to the right nine times (3,200,000,000). The second section contains decimals and fractions, also known as real numbers. Generally, positive numbers that are not whole numbers are given either exactly or to four decimal places as appropriate. The third section contains negative numbers (those less than zero), non-real numbers (those that do not exist on the standard number line) and infinities. Where a given number has an alternative expression, this is displayed in the title.

Collectively, the entries in this book present a rich and varied journey through the world of numbers. One minute you may be reading about a neat solution to an engineering problem, the next puzzling over a deep, unsolved geometry problem. Cute number patterns sit side by side with mindboggling concepts such as infinity, or numbers so large they cannot be expressed even using all the atoms of the universe. Whether you choose to dip in or read from cover to cover, it's a journey full of surprises.

0

Zero is the only whole number that is neither negative nor positive. For a long time, it was not considered to be a number at all, because how can nothing be something?

The word 'zero' derives from the Arabic *sifr*, meaning 'empty', and the earliest known use of a true zero is from an Indian manuscript dating from around 300 CE. In South America, the ancient Mayans independently invented the concept of zero in their base-20 counting system (see page 48). In a positional number system such as this, or the decimal one (see page 28), a zero symbol indicates that a column is empty. Without it, we would not be able to tell the difference between 9, 90 and 900. When it comes to arithmetic, 0 is called the 'additive identity' because any number added to 0 remains unchanged. Any number multiplied by 0 is 0, making zero a multiple of every number – as such, zero is even, because it is a multiple of two. It does not make sense to divide a non-zero number by zero, since no copies of zero will ever equal that number.

1

The number one is the first positive whole number. Every other whole number can be made by adding one to itself so many times. One is also called the multiplicative identity, because any number multiplied by one remains unchanged. This also means that any power of one and any root of one is always equal to one.

One is the only whole number that is neither prime (divisible by two distinct numbers that are one and itself) nor composite (divisible by more than two numbers), though it was considered prime for much of history. The decision to exclude it from being prime was to allow primes to be considered the unique building blocks of all other whole numbers, where each number can be expressed in one way only as the product of primes.

There is no other number that can generate primes by repeating the same digit. For example, 11 is prime, as is 1111111111111111111 (with nineteen digits). The largest known example has 270,343 digits and it is conjectured that there are infinitely many more.

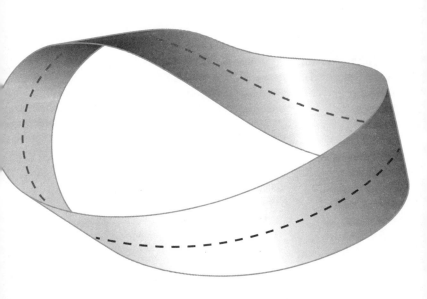

It is possible to create a one-sided shape by gluing together two ends of a strip of paper with a half twist. Such a shape is called a Möbius strip. Cutting it in half along the dotted line leaves, magically, one single loop.

2

—

Our symbol for the number two derives from drawing two horizontal lines in one continuous motion, without raising pen from paper. The smallest positive even number, two is also the only even prime number (which in some sense makes it very odd) and the only prime without an 'e' in its name.

With two as its base, binary arithmetic is the counting system used by computers. It has just two digits: 0 and 1. Each successive column to the left multiplies the value of the digit by two, just as they are multiplied by ten in a decimal system. So, in binary, 1 is one, 10 is two, 100 is four, 1000 is eight, and so on. The figure 1011 means $8 + 2 + 1 = 11$. Computers can easily represent numbers in this way since the 0/1 binary digit is the on/off of a switch.

This number is important for life, since DNA is formed of two strands that wind around each other in a double helix. Almost all animals have bilateral symmetry, meaning they can be divided into two halves that are mirror images of each other.

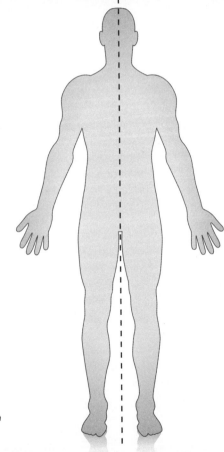

Humans are among the animals that have bilateral symmetry.

3

The first odd prime number, three is also the first Fermat prime and the first Mersenne prime (see pages 42 and 70). Our symbol for the number derives from drawing three horizontal lines in one continuous motion, without raising pen from paper. To tell if a number is divisible by three, add up its digits and see if the answer is a multiple of three. For example, 528 is divisible by three, because 5 + 2 + 8 = 15, which is a multiple of three.

Of all concepts associated with the number three, one of the most familiar is the triangle. This is a very important shape, since any other polygon – a two-dimensional shape made with straight lines – can be built using triangles. Furthermore, three of the five Platonic solids are built using equilateral triangular faces.

A triangular number is one that is the sum of the first *n* whole numbers and is so-called because it can be drawn as dots arranged in a triangle (see page 29). The number three is itself triangular, since it is equal to 1 + 2.

Platonic solids

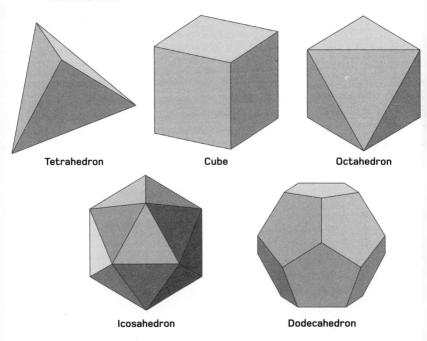

Tetrahedron

Cube

Octahedron

Icosahedron

Dodecahedron

The Platonic solids are symmetric three-dimensional shapes.
Each one is made up of faces that are identical regular polygons.
The tetrahedron, octahedron and icosahedron are built using
equilateral triangular faces.

4

—

$\mathbf{B}$eing equal to 2 + 2 and 2 × 2, the number four is the only number to be both the sum and the product of the same two whole numbers. Four is also the first composite number and the first square number. Also, every whole number can be written as the sum of four square numbers. For example, $34 = 5^2 + 2^2 + 2^2 + 1^2$.

The symbol for four originally consisted of four lines coming together at a point, like a plus sign. It is associated with the Cartesian plane, since the plane is divided (by a cross) into four quadrants. This is also what gives us the four cardinal directions (north, south, east, west).

A flat shape with four sides is called a quadrilateral – for example, a square or a rectangle. A three-dimensional shape with four faces is called a tetrahedron. The tetrahedron is the simplest polyhedron and is made of four triangular faces, six edges and four vertices (corners). When the triangles are equilateral, this produces the regular tetrahedron, the simplest Platonic solid (see page 15).

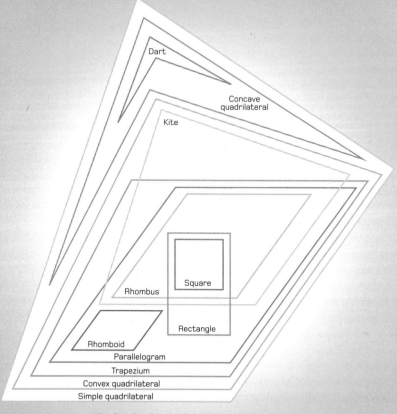

The family of quadrilaterals includes squares,
rectangles, trapeziums, rhombuses and parallelograms.

5

The prime number five is also a Fermat prime, being equal to $2^{2^1} + 1$ (see page 42). It is the only number to be a twin prime in two ways – that is, a prime separated from the next prime by two (3 and 5; 5 and 7). Being separated from the next prime (11) by six, five is also the first sexy prime (see page 66). This number is important in counting due to humans having five fingers on each hand. Many cultures count in groups of five or ten, or have a place-value system that uses a multiple of five.

Five-sided polygons are called pentagons. Pentagons are the only shapes to have the same number of sides as diagonals. Twelve pentagons can be used to build a dodecahedron, one of the five Platonic solids (see page 15).

Draw five dots on a piece of paper and join every pair of dots with a line (straight or curvy). It is impossible to do this without at least two of the lines crossing over each other. This is the smallest such graph with this property (see opposite).

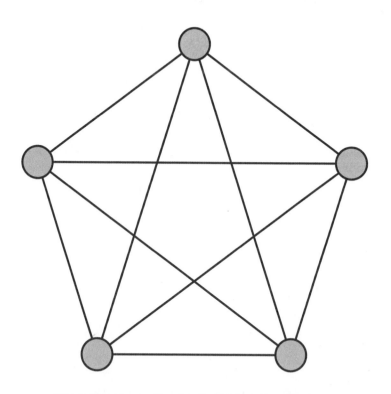

This graph on five vertices is called K5. In mathematical terms, a graph that cannot be drawn without its lines crossing is described as being 'not planar'.

6

The second-smallest composite number, six is equal to 2×3. It is also a triangular number, since it is equal to $1 + 2 + 3$. The numbers one, two and three also happen to be all the numbers that divide exactly into six (excluding itself) – they are called the proper divisors of six. Whenever the proper divisors of a number sum to itself, the number is called perfect. Six is the smallest perfect number.

The Borromean rings pictured opposite have six crossings and are the simplest Brunnian link. This means that the rings are interlocked, but removing any one of the three rings releases the others. Brunnian links can be constructed using any number of interlocking loops.

It is possible to arrange six circles of the same diameter so that they each 'kiss', or touch, a central circle, also of the same diameter. However, it is impossible to do this with seven circles. This makes six the kissing number in two dimensions.

Borromean rings

7

Not only is seven a prime number, it is the only prime to be one less than a perfect cube – that is, the cube of a whole number. Also, being equal to $2^3 - 1$, it is a Mersenne prime (one less than a power of two; see also, page 70).

On a regular six-sided die, opposite numbers always add up to seven. When throwing two dice together, seven is the most common total, since it has the most combinations of numbers that sum to it ($6 + 1$, $5 + 2$, $4 + 3$, $3 + 4$, $2 + 5$, $1 + 6$).

In 1736, Swiss mathematician Leonhard Euler examined a problem called the Seven Bridges of Königsberg. Residents of the city wanted to know if it was possible to cross each of its bridges just once in one continuous walk. Although Euler discovered that it couldn't be done, his visualization of the seven bridges as a network in order to reach his conclusion became the basis of graph theory, an important branch of mathematics with applications in various disciplines today, including computer science.

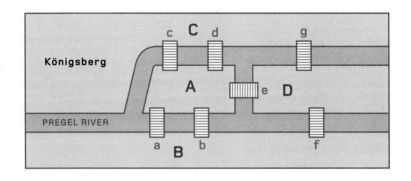

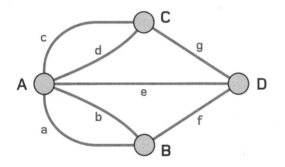

A diagrammatic sketch of Königsberg with its seven bridges (top), and Euler's rendition of the arrangement as a network (left).

8

A fter one, the number eight is the first cube number, as it is equal to 2 × 2 × 2, or 2^3. This means that eight is the volume of a cube where each side has a length of two.

In three-dimensional geometry, a regular octahedron is a shape made of eight equilateral triangles, with four triangles meeting at each vertex (corner). The shape appears as two square-based pyramids placed base-to-base. Connecting the centre point of each face with those of adjacent faces creates the dual shape of the octahedron, which is a cube. In this way the eight faces of the octahedron give rise to the eight vertices of the cube, and the six faces of the cube give rise to the six vertices of the octahedron (see opposite).

Eight is also a Fibonacci number (see page 50). It is the only Fibonacci number to be a perfect cube, and the last Fibonacci number to be one more than a prime.

Duality

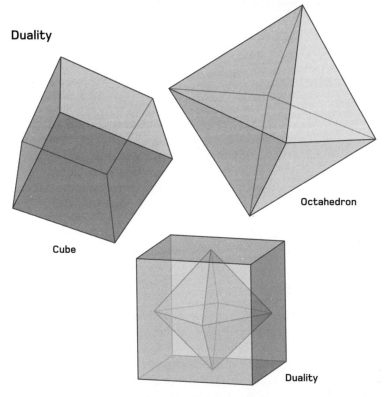

Cube

Octahedron

Duality

When drawing line segments between the centre points of a regular polygon's adjacent faces, those line segments become the edges of that polygon's dual. Each regular polygon has a dual.

9

The last single-digit number in our base-10 decimal system, nine is a square number, being equal to 3 × 3. It is also the first odd number to be composite instead of prime. The number nine is the only integer (positive or negative whole number) power that is one more than another integer power: $3^2 = 1 + 2^3$. This was conjectured by Eugène Catalan in 1844, but not proved until 2002, by Preda Mihăilescu.

To tell if a number is a multiple of nine, add its digits together and see if the answer can be divided by nine. If the answer is still too big to decide, add the digits of the answer, and so on. For example, 8757 is a multiple of 9 since 8 + 7 + 5 + 7 = 27, and 2 + 7 = 9.

You can calculate the nine times table using the fingers on two hands. To find the answer to n × 9, hold your hands out in front of you and curl down the nth finger. The number of fingers (and thumbs!) to the left of the curled finger provides the first digit of the answer, and the number of fingers and thumbs to the right gives you the second digit.

Nine times table

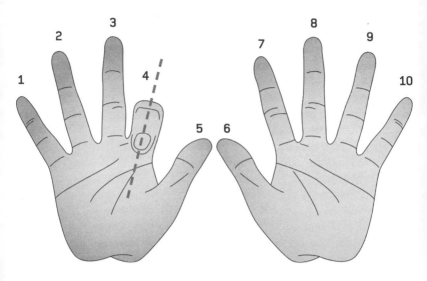

To calculate 4 × 9, hold your hands out in front of you and curl down the fourth finger. There are three fingers to the left of the curled finger, and six to the right, so the answer is 36.

10

Ten is the base for the decimal number system – the standard system used for denoting numbers around the world. Decimal uses ten digits, 0 to 9, and a place-value system, where the value of a digit depends on its place within the number. Each successive column to the left of the decimal place multiplies a digit by ten, and each column to the right divides the digit by ten.

For example:
492.45 means $(4 \times 100) + (9 \times 10) + (2 \times 1) + (4 \times \frac{1}{10}) + (5 \times \frac{1}{100})$.

Many societies around the world independently chose ten as their number base. This likely happened because people started counting on their fingers and we have ten fingers. Our current notation for numbers is derived from the Hindu-Arabic numeral system. It became widely used because it makes it easy to multiply and divide large numbers, as compared with additive number systems such as Roman numerals.

Triangular numbers

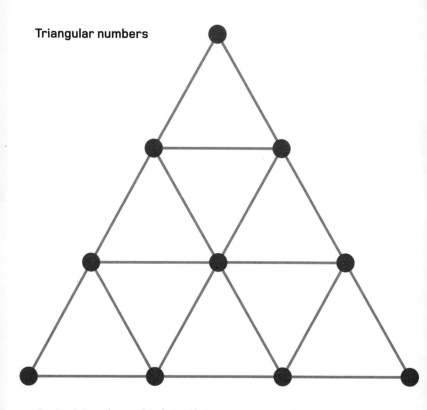

Ten is a triangular number, being the sum 1 + 2 + 3 + 4. The mathematical name for this diagram is 'tractys'.

11

Eleven is a prime number that is also the first repdigit in the decimal number system. This means it is composed of the same digit repeated a number of times. The numbers 222 and 9999 are also examples of repdigits.

To test whether a number is divisible by eleven, successively add and subtract its digits. If the answer is a multiple of eleven, then so is the original number. For example, 8162 is a multiple of eleven since $8 - 1 + 6 - 2 = 11$. To multiply a two-digit number by eleven, add the two digits together and place the answer in between the originals. So, $34 \times 11 = 3(3 + 4)4 = 374$. Sometimes you may need to carry a ten, so $39 \times 11 = 3(12)9 = 429$.

In many languages eleven is the first compound number – it's expressed in terms of more than one unit. For example, in Hungarian eleven is *tizenegy*, meaning 'one on ten'. In Germanic languages, including English, eleven and German *elf* derive from the Proto-Germanic *ainalif* meaning 'one left (over after ten)'.

Sporting numbers

Eleven is often the number of players on a sports team. Cricket, soccer, American football and field hockey all have teams with eleven players.

12

Twelve is a composite number equal to 2 × 2 × 3. It has six divisors (1, 2, 3, 4, 6, 12). This is more than any smaller number, which means twelve can be described as a highly composite number. The fact that it has so many divisors makes twelve an attractive base to use as a number system. Fractions are easier to calculate in base-12 (duodecimal) than in decimal. Pre-metric systems often have aspects of duodecimal, with twelve inches in a foot and twelve pence in a shilling. The number is also a feature of time, with a day being two sets of twelve hours, and a year being twelve months.

Almost all of the five Platonic solids exhibit twelve-ness: cubes and octahedra have twelve edges, dodecahedra have twelve faces and icosahedra have twelve vertices. In three-dimensional geometry, it is possible for twelve equal-sized spheres to touch a central sphere of the same size simultaneously. The result was not proven until 1953, since it was unclear whether there might be a clever way to fit in a thirteenth sphere.

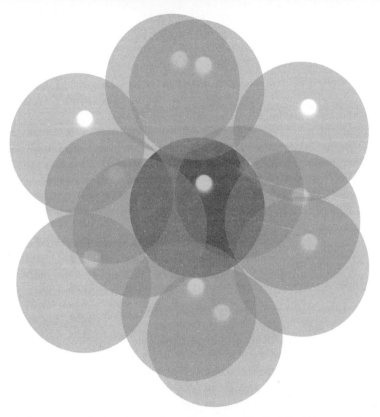

Twelve spheres clustered around a central sphere. The fact that they can all touch the central sphere makes twelve the kissing number in three dimensions.

13

Fear of the number thirteen is called triskaidekaphobia. Superstition around this number is common in several cultures, though the number is considered lucky by others. A prime number, thirteen is also the smallest emirp, which means it is still prime when its digits are reversed (13 – 31).

There are thirteen Archimedean solids. These are convex three-dimensional shapes with faces that are regular polygons and all vertices identical. This means you can rotate the shape so that all vertices line up exactly with each other. They differ from Platonic solids in that their faces need not all be the same shape. An example is the icosidodecahedron, which is built from pentagons and equilateral triangles. At each vertex, two pentagons and two triangles meet, always in the same configuration. The thirteen Catalan solids are the duals of the Archimedean solids. This means each face is replaced with a vertex and each vertex with a face. The faces of Catalan solids are identical, but they are not regular polygons.

Archimedean and Catalan solids

An Archimedean icosidodecahedron (left) and its dual, the
Catalan rhombic triacontahedron (right).

14

P ictured opposite are two mathematical puzzles that both, coincidentally, have an answer of fourteen. The first is finding the number of ways a hexagon can be divided into four triangles using non-overlapping lines. The second is counting the paths in a 4 × 4 grid that travel from the bottom-left corner to the top-right corner, moving rightwards or upwards, but never crossing the diagonal. In fact, the seeming coincidence of the answers is no coincidence at all. The number of ways of dividing an $(n + 2)$-sided shape into n triangles is the same as the number of paths in an $n \times n$ grid that do not cross the diagonal. The sequence of numbers created this way are called Catalan numbers; 14 is the Catalan number for $n = 4$. The formula for the n-th Catalan number is $\frac{1}{n+1}\binom{2n}{n}$ where $\binom{2n}{n}$ is the number of ways of choosing n objects out of $2n$ (see page 78). The numbers have popped up as solutions to more than sixty different problems in mathematics, and are especially important in computer science.

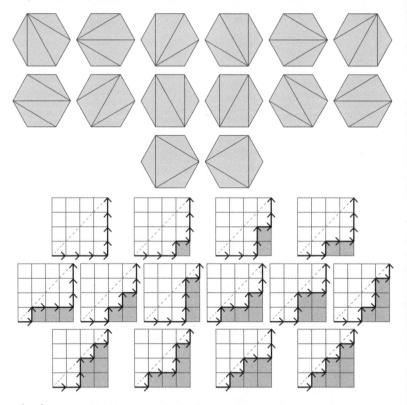

(Top) There are fourteen ways to divide a hexagon into four triangles.
(Bottom) There are fourteen paths from corner to corner in a 4 × 4 grid that never cross the diagonal and only move rightwards and upwards.

15

Fifteen is a semiprime, being the product of the two primes three and five. It is also the smallest emirpimes (semiprime spelt backwards), as both it and its reversal (51) are semiprimes. Being the sum of the numbers one to five, fifteen is also a triangular number (see page 29). It is possible to arrange the digits 1 to 9 in a three by three grid so that every row, column and diagonal adds up to the same number. This arrangement is called a magic square, and its magic total is fifteen. There is only one way to solve this puzzle, known to Chinese mathematicians as early as 190 BCE.

It is not possible to tile the plane with regular pentagons – that is, to tessellate the pentagons so that they fill the space precisely, without overlaps or gaps. This is because the interior angle of a pentagon, 108°, does not divide into 360. Several types of irregular pentagon can be tiled, however. There are fifteen known types of convex pentagon that tile the plane, with the most recent only discovered in 2015.

Tiling the plane

In 2017, Michaël Rao submitted a proof that only 15 types of convex pentagon tile the plane, but this is still being checked.

16

The smallest number after one to qualify as being a fourth power, sixteen is equal to 2^4, or $2 \times 2 \times 2 \times 2$. It can be written as both 2^4 and 4^2 and is the only number with this property – that is, to be equal to a^b and b^a where the values of a and b are different.

A four-dimensional cube is called a tesseract. Just as a cube can be thought of as six squares assembled to make a three-dimensional shape, a tesseract can be thought of as eight cubes assembled to make a four-dimensional shape. It is one of six regular polytopes in four dimensions, analogous to the Platonic solids in three dimensions. The term 'polytope' is used to describe a geometric shape with flat sides. A tesseract has sixteen vertices. These sixteen corners are located at the points in four-dimensional space with coordinates $(\pm1, \pm1, \pm1, \pm1)$. This is similar to how a square in two dimensions has its four corners at the coordinates $(1, 1)$, $(1, -1)$, $(-1, 1)$ and $(-1, -1)$.

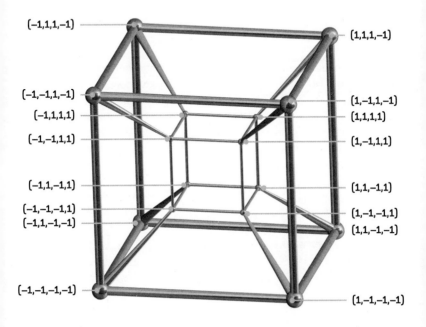

(−1,1,1,−1) ─── (1,1,1,−1)

(−1,−1,1,−1) ─── (1,−1,1,−1)
(−1,1,1,1) ─── (1,1,1,1)
(−1,−1,1,1) ─── (1,−1,1,1)

(−1,1,−1,1) ─── (1,1,−1,1)
(−1,−1,−1,1) ─── (1,−1,−1,1)
(−1,1,−1,−1) ─── (1,1,−1,−1)

(−1,−1,−1,−1) ─── (1,−1,−1,−1)

Projection into three dimensions of a tesseract, a four-dimensional cube. The
four coordinates represent left/right, front/back, down/up and outer/inner.

17

Since seventeen is equal to $2^3 + 3^2$, it is the only prime that can be written as $p^q + q^p$ for primes p and q. It is also the only prime number that is the sum of four consecutive primes (2 + 3 + 5 + 7). Seventeen is a twin prime (with nineteen) and a Fermat prime (one more than a power of two, where the power is itself a power of two) – in this case $2^{2^2} + 1$. To date, only five Fermat primes have been identified: 3, 5, 17, 257 and 65537.

There are exactly seventeen different symmetries that a wallpaper pattern can have, in terms of the ways in which a motif can be translated, reflected and rotated. These are called the plane symmetry groups. For example, a wallpaper pattern showing 'p31m' symmetry has order-3 rotations (that is, around 120°), reflections, and glides (reflection plus translation). Two wallpapers with the same collection of symmetries are considered to be the same by mathematicians, even if the colours or motifs differ. In this way there are only seventeen types of wallpaper pattern.

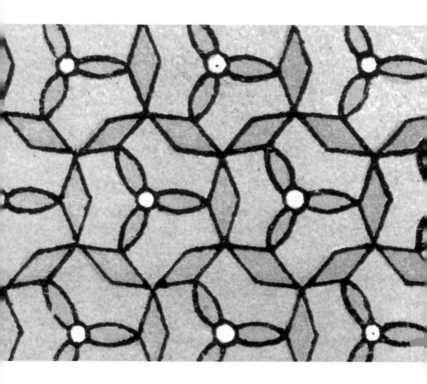

A Persian glazed tile showing 'p31m' symmetry.

18

A composite number equal to 2 × 3 × 3, eighteen is also semiperfect, since it is equal to the sum of some of its divisors (18 = 3 + 6 + 9). An emirp is a prime whose reversal is also prime – for example, 37 (with reversal 73). If you take any emirp and subtract its reversal, the answer is always a multiple of eighteen.

In recreational mathematics there are eighteen different pentominoes. These are all the shapes that can be made using five squares, including mirror reflections. First described by Solomon Golomb in 1965, the name is a combination of 'penta' and 'domino'. Each of the eighteen pentominoes can tile the plane (see page 38). The twelve unique pentominoes – those excluding mirrors – can fit together to form various rectangles, with sizes 6 × 10, 5 × 12, 4 × 15 and 3 × 20. These are often sold as puzzle games, an additional challenge being to work out how many solutions there are for each size. The computer game Tetris was inspired by pentominoes and uses tetrominoes (made of four squares), as pentominoes were considered to be too difficult.

The eighteen pentominoes

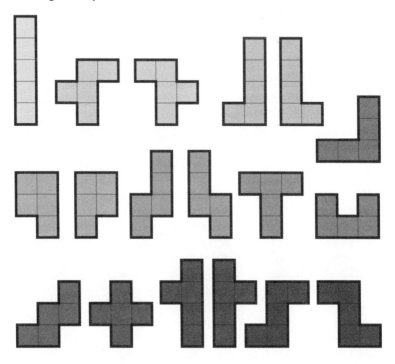

19

This prime number has several curious properties: it is the only prime equal to the difference between the cubes of primes $(3^3 - 2^3)$; it is the smallest prime that can be turned upside down to make a different prime (61); and it is the largest prime palindrome in Roman numerals (XIX). Raising nineteen to its own power gives a number that is pandigital – that is, 19^{19} contains all the digits from 0 to 9. The number nineteen is the smallest number with this property.

Nineteen is important in calendar systems, since nineteen solar years (a Metonic cycle) is very close to being a whole number of lunar months. Babylonian and Hebrew calendars were based on twelve lunar months per year, with a thirteenth month added seven times over each nineteen-year period. The Metonic cycle is still used for calculating the date of Easter. Another potential candidate for a lunar calendar, the Ishango bone pictured opposite contains groups of notches on three sides, with the number nineteen represented twice.

The Ishango bone

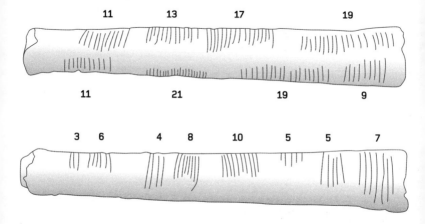

One of the oldest discovered mathematical objects, the Ishango bone was found on the border of Uganda and the Democratic Republic of Congo, and is likely more than 20,000 years old. Opinions differ as to whether it could be a table of primes, a calculating tool or a lunar calendar.

20

Twenty is a composite number equal to 2 × 2 × 5. The word 'score' comes from the old English word for twenty and is still sometimes used to refer to groups of twenty items. In geometry, the icosahedron is a three-dimensional shape with twenty faces. Its dual, with twenty vertices, is the dodecahedron (see page 24).

A vigesimal number system is one that uses twenty as its base. Such systems need twenty digit-symbols for the numbers zero to nineteen. The placement of digits in columns indicates their value, with each column worth twenty times more than the column to its right. Vigesimal systems have arisen around the world, most likely from people using their fingers and toes to count. They are found today in Bhutan, India and Africa, and exist in Celtic languages as well as in modern French (where the term for eighty, *quatre-vingts*, literally means 'four twenties'). Most famously it was the number base used by the ancient Mayans and Aztecs in South America.

Mayan numerals

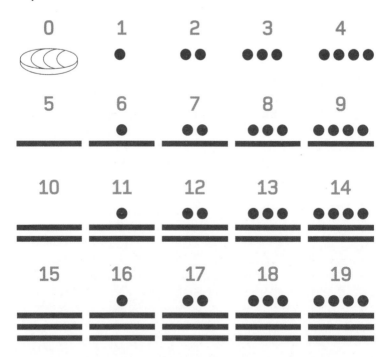

The digit-symbols for Mayan numbers zero to nineteen as used in their base-20 number system.

21

Can you find the next number in the following sequence: 1, 1, 2, 3, 5, 8, 13, ...? Called the Fibonacci sequence, each number is calculated by adding the two previous numbers together. The answer to this riddle is 8 + 13 = 21. Of all the Fibonacci numbers, twenty-one is extra special. Not only is it a Fibonacci number itself but so, too, are each of its digits (2 and 1) and the sum of those digits (2 + 1 = 3).

Fibonacci numbers date back to India, in 200 BCE, and were introduced to Europe in 1202 by Leonardo of Pisa (also called Fibonacci). In his book the *Liber Abaci*, Fibonacci studies the growth of a rabbit population. It starts with one breeding pair. The idea is that a pair of rabbits produces a new pair of babies every month, which then grow and follow the same pattern. Fibonacci calculated that the number of rabbits produced each month follows the Fibonacci sequence. Fibonacci numbers relate to the golden ratio (see page 328) and to the arrangements of spirals in plants and sunflower seeds (see page 386).

Fibonacci's bunnies

In Fibonacci's model, an adult pair of rabbits produces a pair of offspring each month, and a baby pair of rabbits takes one month to mature to adulthood.

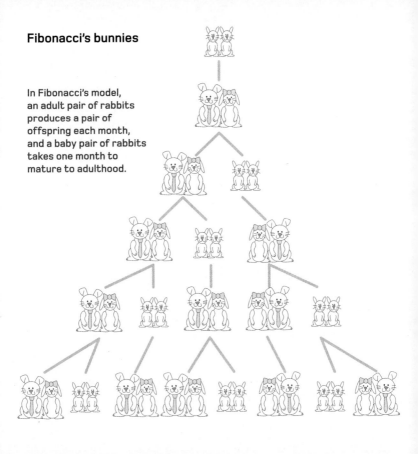

22

I have a bar of chocolate that is eight squares long. How many different ways can I break it up into smaller squares? I can break it in half to get a 4 + 4 split. Or I can break off three single squares to get a 5 + 1 + 1 + 1 split. Can you find them all? (In this problem a 5 + 1 + 1 + 1 split is the same as a 1 + 1 + 5 + 1 split.)

Splitting one whole number into a collection of smaller whole numbers is called a partition. In this case, there are twenty-two ways to partition the number eight. The ability to find all partitions of a number is important for many areas of maths and science, including quantum systems, molecular chemistry, genetics and statistical physics. Yet, surprisingly, for such a simple problem, nobody has found a basic formula for predicting how many partitions a number has. Instead, there are approximate formulae that become more accurate as a number gets bigger, or formulae that depend on knowing the answers for all smaller numbers. But no straightforward formula has been arrived at to date.

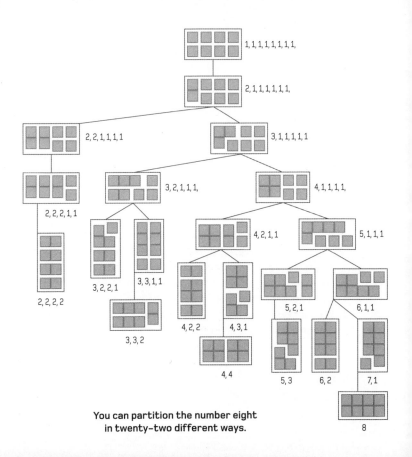

You can partition the number eight
in twenty-two different ways.

23

In my class of thirty students at school, two shared the same birthday. This might seem like a big coincidence, given that there are 365 birthdays in the year (ignoring leap years). In fact, there is a 70% chance of this happening. So how many people need to be in a room before it is more likely than not that two share a birthday? The answer is twenty-three, in a result called the birthday paradox. The calculation is worked out by solving the easier problem of the chances of nobody sharing a birthday out of twenty-three people. Suppose the people walk into the room one by one. Person two has 364 out of 365 birthdays to choose from to avoid matching person one. Then person three has 363 out of 365 birthdays to avoid matching either of the first two, and so on. When person twenty-three walks in, they have 343 out of 365 birthdays to choose from. Multiplying these together gives us:

$$\frac{364}{365} \times \frac{363}{365} \times \frac{362}{365} \times ... \times \frac{343}{365}$$

which is about 49%. This means the chances of at least two people sharing a birthday is about 51%.

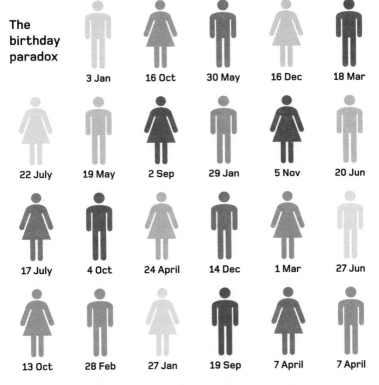

The birthday paradox

3 Jan   16 Oct   30 May   16 Dec   18 Mar

22 July   19 May   2 Sep   29 Jan   5 Nov   20 Jun

17 July   4 Oct   24 April   14 Dec   1 Mar   27 Jun

13 Oct   28 Feb   27 Jan   19 Sep   7 April   7 April

How many people need to be in a room before there is more than a 50% chance that two share the same birthday? Although there are 365 days in a year, the answer is a lot smaller than 182!

24

A factorial number, written $n!$, is the number you get by multiplying all the whole numbers between one and n. For example, $4! = 1 \times 2 \times 3 \times 4 = 24$. Factorial numbers are important in an area of mathematics called combinatorics, which measures how many ways there are of permuting or combining objects. The number $4! = 24$ is the number of ways of arranging four objects, arrived at because there are four choices for the first object, then three remaining choices for the second object, two choices for the third object, and only one choice for the fourth object.

Besides combinatorics, twenty-four also has strong connections to prime numbers. Take any two prime numbers bigger than three and square them. For example, $7^2 = 49$, and $13^2 = 169$. The difference between these two numbers (in this case, $169 - 49 = 120$) will always be a multiple of twenty-four. You can also square any prime bigger than three, subtract one, and your answer will be a multiple of twenty-four. For example, $19^2 - 1 = 360 = 24 \times 15$.

Combinatorics

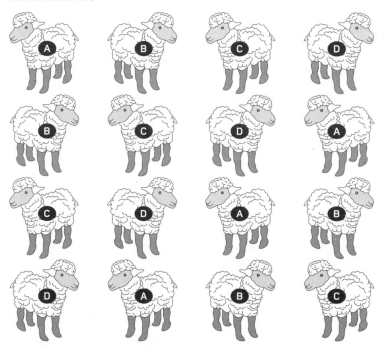

There are twenty-four ways of arranging four sheep in line.
Called A, B, C and D, you could have ABCD, BCDA, CDAB, DABC, and so on.
Can you find all the permutations?

25

Pythagoras' theorem says that, in a right-angled triangle with short sides of length a and b, and the longest side of length c, then $c^2 = a^2 + b^2$. In words, the square of the hypotenuse (the longest side) is equal to the sum of the squares of the other two sides. The most famous example of this is the 3:4:5 triangle, where you have $5^2 = 3^2 + 4^2$, or $25 = 9 + 16$. Since 25 is the smallest square number that is the sum of two other (non-zero) square numbers, the 3:4:5 triangle is the smallest right-angled triangle with whole-number sides.

The opposite of Pythagoras' theorem is also true: if you can find three numbers a, b and c so that $c^2 = a^2 + b^2$, then a, b and c will form the sides of a right-angled triangle. Historians have wondered whether the ancient Egyptians used this method to measure accurate right angles. For example, knowing that $25 = 9 + 16$ would have enabled them to use a rope with knots tied at regular intervals to lay out a 3:4:5 triangle and get a right angle.

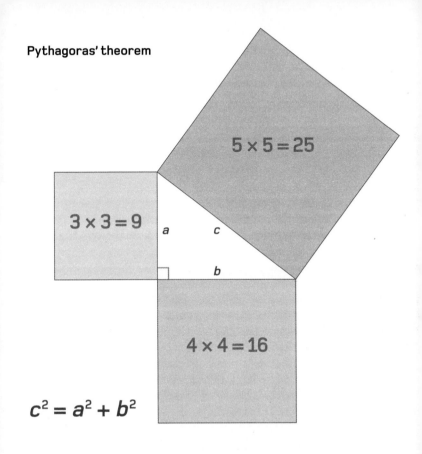

Pythagoras' theorem

$3 \times 3 = 9$

$5 \times 5 = 25$

a   *c*

b

$4 \times 4 = 16$

$c^2 = a^2 + b^2$

26

Twenty-six is the only number to be simultaneously one more than a square number ($26 = 5 \times 5 + 1$) and one less than a cube number ($26 = 3 \times 3 \times 3 - 1$).

According to a mathematical proof called the 'classification of finite simple groups', which runs to tens of thousands of pages, twenty-six is the number of what mathematicians call sporadic finite simple groups. Groups are mathematical objects that capture the structure of symmetry. For example, a square's symmetries are different from those of a human face, and a cylinder's symmetries are different from those of a strand of DNA. There are infinitely many different groups, but they are all built from the same building blocks, called simple groups. In 2004, mathematicians succeeded in showing that every finite simple group either belongs to one of three well-behaved families, or is one of twenty-six sporadic groups that do not follow any pattern. The largest of these is called the Monster group (see page 240).

Symmetry

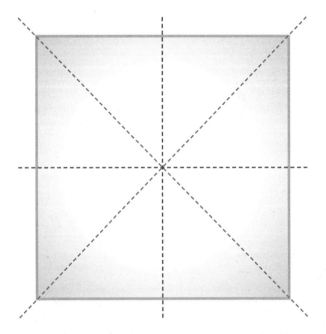

A symmetry of a shape is any action that leaves it looking the same.
For example, the symmetries of a square consist of four rotations
(by 0°, 90°, 180° and 270°) and four reflections (horizontally,
vertically and through the two diagonals).

27

The cube of a number, n, is three instances of that number multiplied: $n \times n \times n$, also written as n^3 (n to the power of three). For example, twenty-seven is the cube of three because $27 = 3 \times 3 \times 3$. A cube number n^3 is so named because it is the volume of a cube with side lengths n.

If someone hands you a random number, it is not easy to tell immediately if it is a cube, since any of the digits 0 to 9 can appear as the final digit of a cube number. However, there is one test that can quickly check if a number is not a cube. All cubes will have a digital root of 0, 8 or 9. This means that if you add up the digits of the number (doing so repeatedly if you get a multi-digit answer), the answer will be 0, 8 or 9. For example, with twenty-seven you have $2 + 7 = 9$. If you do this trick with 491, you get $4 + 9 + 1 = 14$, and again $1 + 4 = 5$, so 491 cannot be a cube. Take care, however: just because a number's digital root is 0, 8 or 9, doesn't mean the number is necessarily a cube. For example, 17 (digital root 8) is not a cube.

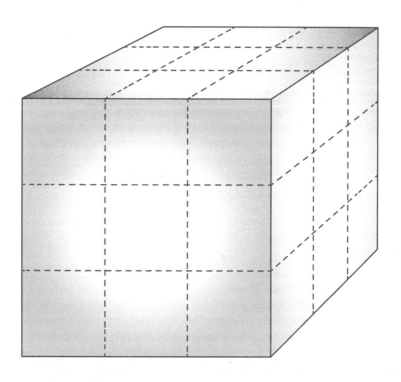

Twenty-seven small cubes can be arranged in a large 3 × 3 × 3 cube.
This is why twenty-seven is called a cube number.

28

A perfect number is one that is the sum of its proper divisors – that is, the sum of all the numbers that divide into it exactly, except itself. Twenty-eight is a perfect number because $28 = 1 + 2 + 4 + 7 + 14$.

Perfect numbers have been investigated since ancient Greek times, appearing in Euclid's *Elements* c. 300 BCE. Euclid found a connection between perfect numbers and special types of primes. He showed that if $2^p - 1$ were prime, then $2^{p-1}(2^p - 1)$ would be perfect. For example, $2^3 - 1 = 7$ is prime, so $2^2(2^3 - 1) = 4 \times 7 = 28$ is perfect.

Twenty-eight is also a triangular number, being the sum of all the numbers from one to seven. This follows from its perfection, since every even perfect number is also a triangular number. More surprisingly, twenty-eight is the sum of the first five primes $(2 + 3 + 5 + 7 + 11)$ and the first five non-primes $(1 + 4 + 6 + 8 + 9)$. The next such number is not until 71208.

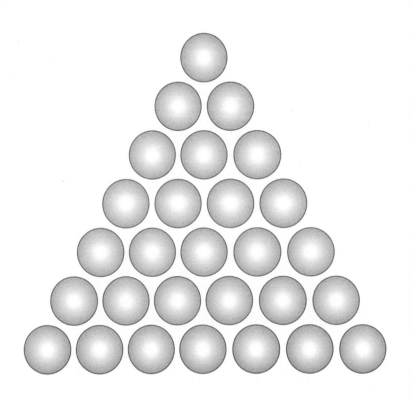

Twenty-eight drawn as a triangular number.

29

Sexy primes are two primes that differ from each other by six. For example, twenty-three and twenty-nine are sexy primes. The name comes from a play on words from the Latin word for six, which is *sex*. Arguably the sexiest prime is twenty-nine, since it is also part of a sexy triplet (17, 23, 29) as well as a sexy quadruplet (11, 17, 23, 29) and a sexy quintuplet (5, 11, 17, 23, 29). There are no other sexy quintuplets, since any five numbers created by adding six each time will contain a multiple of five. Twenty-nine is also a twin prime, since it differs by two from the prime thirty-one. It is unknown if there are infinitely many pairs of twin primes or sexy primes (see page 216).

Try this for an exercise: twenty-nine is the smallest positive whole number that cannot be made using the numbers 1, 2, 3 and 4 exactly once each using the operations addition, subtraction, multiplication and/or division. See if you can work out solutions for numbers one to twenty-eight.

Sexy quadruplet

Sexy prime

5   11   17   23   29

Sexy triplet

Sexy quintuplet

30

The smallest number to be the product of three primes, thirty is equal to $2 \times 3 \times 5$. All numbers smaller than thirty that are coprime to it (that is, they share no common factor) are themselves prime, and thirty is the largest number with this property.

Draw a 4×4 grid and count how many squares there are. If your instinct is to answer sixteen, think again! There are indeed sixteen 1×1 squares, but there are also nine 2×2 squares, four 3×3 squares, and the one big 4×4 square. In total this makes $16 + 9 + 4 + 1 = 30$.

Notice that the above sum can also be written as $4^2 + 3^2 + 2^2 + 1^2$. Numbers that can be written as the sum of the first n squares are called square pyramidal numbers, since they count the number of spheres that can be stacked in a square-based pyramid. The first four square pyramidal numbers are 1, 5, 14 and 30 (see opposite).

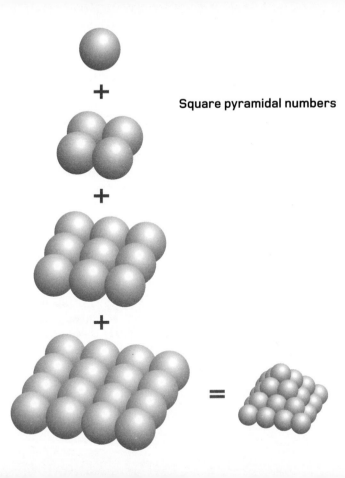

Square pyramidal numbers

31

Separated from another prime (twenty-nine) by two, thirty-one is a twin prime. It is also a Mersenne prime, being one less than a power of two: $31 = 2^5 - 1$. It is the only Mersenne prime where reversing the digits creates another prime.

In IQ tests, it is common to see problems such as 'What comes next in the sequence of numbers?' These are not always fair because any canny mathematician knows that there can be multiple different correct answers to the question.

Take Moser's circle problem. This asks how many regions are created in a circle by drawing n dots around the edge and connecting every pair of dots with a line. (No three of these lines should meet at the same point.) The answers to this problem, starting from $n = 1$ upwards, gives the sequence of numbers 1, 2, 4, 8, 16. A natural guess is that the answers multiply by two each time, so the next answer will be thirty-two. Yet the true answer is thirty-one, making the pattern much more difficult to spot.

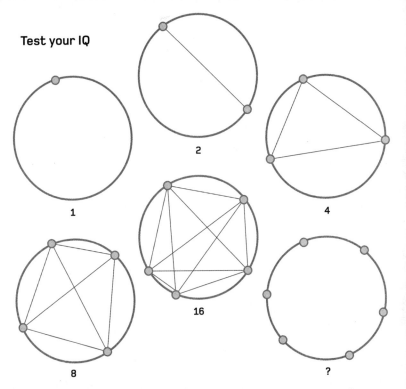

Test your IQ

1

2

4

8

16

?

Draw *n* dots around the edge of a circle, not necessarily equally spaced. Now connect every pair of dots with a line. How many areas do these lines divide the circle into? The answers are given here for one, two, three, four and five dots – can you get the answer for six dots and discover the pattern?

32

Mathematician and comedian Matt Parker has launched a campaign to correct an egregious mathematical error on British road signs. The error is on the symbol for a soccer ground, which depicts soccer balls made up only of hexagons. However, if you try to join a group of hexagons, you will find that you can only produce flat shapes, not curved ones. A true football has thirty-two faces: twelve pentagons and twenty hexagons. Mathematically this shape is called a 'truncated icosahedron', as it is created from an icosahedron (a three-dimensional shape made from twenty equilateral triangles) by slicing off each of its twelve corners. Truncated icosahedra are also known as 'buckyballs', as they are the shape of the molecule buckminsterfullerene, made from sixty carbon molecules (one at each vertex). The discovery of this structure in the 1980s led to its creators winning the Nobel Prize in Chemistry in 1996. There are infinitely many sphere-like shapes made from pentagons and hexagons. However, every single one of these will contain exactly twelve pentagons.

The British road sign for a soccer ground (left) and an actual soccer ball, made of both pentagons and hexagons (right).

33

A deceptively simple-sounding question about cube numbers continues to baffle mathematicians, with thirty-three being the most recent number to have been cracked.

A cube number, written as n^3, is a number n multiplied three times: $n \times n \times n$. The puzzling problem is whether any positive whole number can be written as the sum of three whole number cubes. For example, $36 = 1^3 + 2^3 + 3^3 = 1 + 8 + 27$. Negative numbers are allowed, so we can write 1 as $10^3 + 9^3 - 12^3 = 1000 + 729 - 1728$. We can also use numbers more than once, so can write $3 = 1^3 + 1^3 + 1^3$. So far, mathematicians have shown that the problem is impossible if the number has remainder four or five when divided by nine (e.g. 4, 5, 13, 14, 22, 23). They conjecture that every other number has at least one solution, and until recently thirty-three was the smallest number without an answer. The solution was only found in 2019 using computers, with the cube numbers being sixteen digits long! The smallest unsolved case is now forty-two.

$$33 = a^3 + b^3 + c^3$$

Can you find whole numbers (positive or negative)
a, b and c to make this equation true?

34

The number thirty-four is a Fibonacci number (see page 50). It is also a semiprime, being the product of the two primes two and seventeen. Its neighbours thirty-three and thirty-five are also semiprimes, making thirty-four the smallest number whose neighbours have the same number of factors as itself.

Take a look at the 4 × 4 magic square pictured opposite. It places the numbers one to sixteen in such a way that every row, column and diagonal adds up to the same number, called the magic constant. There are 880 different 4 × 4 magic squares (excluding rotations and reflections), and they all have the same magic constant of thirty-four. This magic square is especially magic. The magic constant thirty-four is not only found in the rows, columns and diagonals, but in every 2 × 2 square, in the corners of any 3 × 3 square, in the four corners, in the numbers one step clockwise from the four corners (that is, 3, 8, 14, 9) and also one step anticlockwise. It is also in the off-diagonal corners (2, 8, 9, 15 and 5, 3, 12, 14).

This 4 × 4 magic square is part of an engraving called *Melancholia* by Albrecht Dürer. The final row holds the numbers 1514, the year the work was made.

35

One day Snow White visits her friends, the seven dwarves, with a task that needs three people to help her. How many ways can she choose three helpers from her seven friends?

This type of problem is known as a 'combination' problem. In its most general form, it asks how many ways we can choose k items out of n, where the order of choosing the items is unimportant. Snow White has seven ways of choosing her first helper, six ways of choosing the second, and five ways of choosing the third, giving $7 \times 6 \times 5 = 210$ total. This then needs to be divided by six since there are six ways of ordering her three choices ($= 3 \times 2 \times 1$), and she doesn't care about order. So she has thirty-five possible choices.

This combination problem has led to the number thirty-five being known as '7 choose 3'. It is the same answer as the number of ways of choosing four items out of seven, since choosing four items is the same as not choosing three.

Pascal's triangle

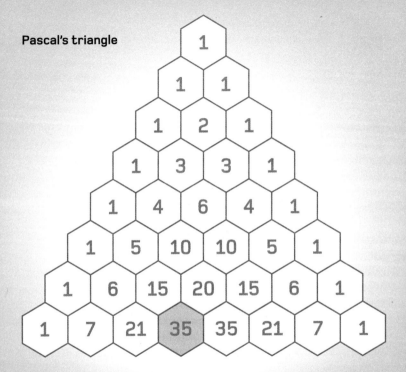

Each number in this grid (called Pascal's triangle) is found by adding the two numbers immediately above it. These numbers solve the problem of how many ways there are of choosing *k* objects out of *n*: look for number (*k* + 1) in row (*n* + 1). So, to find how many ways there are of choosing three objects out of seven, look at the fourth number in the eighth row: 35.

36

Apart from one, thirty-six is the smallest number that is both a square number (equal to 6 × 6) and a triangular number (equal to the sum of the numbers one to eight). For this reason, it is called a square triangular number.

The angle found inside a regular pentagram is 36° – one-tenth of a full rotation. In geometry, the pentagram is an example of a regular star polygon. It is 'regular' because all the lines are the same length, and a 'star polygon' because the lines are allowed to intersect each other. A {p/q} regular star polygon is created by drawing p dots equally spaced around a circle and then drawing lines connecting every pair of points that are q points apart. The regular pentagram is a {5/2} star polygon. It is a fun investigation to explore the different types of star you can draw by changing how many points you start with and which ones you connect. For example, the {6/2} star polygon draws two intersecting triangles, while the {8/3} and {8/5} stars give the same shape. Can you figure out why?

Regular pentagram

37

Think of a number between one and nine, and write it down three times – for example, 111. Now divide this number by the sum of the digits (in this case, $1 + 1 + 1 = 3$, so you calculate $111/3$). Your answer will always be 37!

Fermat's Last Theorem is a conjecture from 1637 that remained unsolved for 350 years. It stated that there could never be three whole numbers x, y and z solving the equation $x^n + y^n = z^n$ for any n bigger than two. In 1847, French mathematician Gabriel Lamé announced a solution. He had factorized Fermat's equation using complex numbers (see page 378), but it turned out he had overlooked a problem. With whole numbers, factorization into primes is unique, but with the types of complex numbers Lamé was using it was not the case. Although mathematicians tried to patch up the proof, a class of primes called irregular primes (of which thirty-seven is the smallest) still stopped it working. Another 147 years passed before Andrew Wiles proved Fermat's Last Theorem by attacking it from a different direction altogether.

Thirty-seven is a centred hexagonal number, meaning it can be represented by a central dot surrounded by hexagonal layers of dots.

38

It is a famous unsolved mathematical conjecture that every even number can be written as the sum of two prime numbers (see page 230). You may well wonder about a sister problem to this one: whether every even number can be written as the sum of two odd composite numbers.

For example, 50 = 15 + 35. Here 15 and 35 are composite because they can each be divided by numbers other than one and themselves. The smallest number that can be written like this is 18 = 9 + 9, since nine is the smallest odd composite number. Breaking down an even number into the sum of two odd composite numbers becomes easier as the numbers get larger, since the primes become scarcer as a proportion of the odd numbers. Out of the first ten odd numbers above one, 70% are prime, but out of the first hundred odd numbers, only 46% are prime. It turns out that thirty-eight is the biggest number that cannot be written as the sum of two composite odd numbers – anything bigger than this will work.

$$38 = ? + ?$$

Can you write thirty-eight as the sum
of two odd composite numbers?

39

In the Penguin *Dictionary of Curious and Interesting Numbers* by David Wells, thirty-nine is listed as being the first uninteresting number. This creates a paradox, since the smallest uninteresting number is itself interesting. Despite the paradox, people have persevered in trying to find uninteresting numbers. Currently, 262 is the smallest number on Wikipedia not to have its own entry, while 17843 is the smallest number not to appear in the On-Line Encyclopedia of Integer Sequences (OEIS) database of interesting sequences. David Wells was, perhaps, overly harsh with thirty-nine. It is the smallest odd semiprime (product of two primes) to be the sum of all the primes between its two factors. In other words, 39 = 3 × 13 and 39 = 3 + 5 + 7 + 11 + 13. This number also has a few connections with the number three: it is the sum of the first three powers of three $(3^1 + 3^2 + 3^3)$ and it is also the smallest number that can be partitioned into three parts in three different ways, with each partition having the same product: {6, 8, 25}, {5, 10, 24} and {4, 15, 20}.

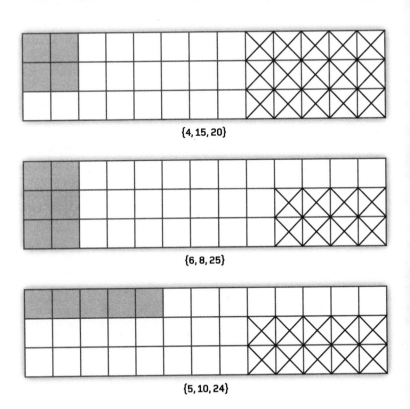

{4, 15, 20}

{6, 8, 25}

{5, 10, 24}

The number thirty-nine partitioned in three ways into three parts.
The product of the numbers in each partition is 1200.

40

There are forty possible two-digit endings for a prime number. A prime cannot end in 35, else it would be divisible by 5, but it could end in 39 (for example, 239 is prime). Among the first 10,000 primes, the ending 57 appears the most often, but after the first billion primes, 47 takes the lead.

Linguistically and scientifically, forty has a few fun facts. It is the only number in the English language whose letters are spelt in alphabetical order. (We can thank our seventeenth-century ancestors who changed the spelling from 'fourty'.) Its friend is the number 'one', which is the only number with its letters in reverse alphabetical order. The word 'quarantine', meaning to keep people in isolation if suspected of having a disease, comes from the Italian *quarantina giorni*, meaning 'forty days'. During the Black Death in Europe in the fourteenth century, the city of Venice enforced a rule saying that ships suspected of carrying plague victims had to wait off port for forty days before being allowed to enter the city.

In terms of temperature, −40 degrees is the only temperature that is the same in both the Celsius and Fahrenheit scales.

41

Patterns in prime numbers are notoriously difficult to find. Many of today's famous unsolved problems in mathematics concern primes: the Goldbach conjecture, the twin-prime conjecture and the Riemann hypothesis, for example (see pages 230, 338 and 378). An obvious question is whether there is a formula for predicting where to find the primes. The answer is 'yes', but the formulae we have found so far are incredibly unwieldy and almost impossible to use in any practical sense.

A polynomial is the simplest kind of mathematical expression. It consists of variables and numbers that are combined using only addition, subtraction and multiplication. An example is $x^2 + 2xy - 5$. So, can a polynomial formula be found that produces only primes? The expression $x^2 - x + 41$ is a polynomial that produces primes from $x = 0$ up to 40, the longest possible run. It was discovered by Leonhard Euler in 1771, and forty-one is thus called a 'lucky number of Euler'. Sadly it has been shown that no polynomial will ever produce only primes.

Ulam spiral

Stanislaw Ulam was doodling in a meeting one day, writing the whole numbers 1, 2, 3, 4, ... round in a spiral. Highlighting the prime numbers, he found that they were not random but appeared to form straight lines. These correspond to quadratic polynomials such as $x^2 - x + 41$, which produce an unusually large number of primes. The reason for this is still being investigated.

42

Besides being the answer to the Ultimate Question of Life, the Universe and Everything (according to *Hitchhiker's Guide to the Galaxy* author Douglas Adams), forty-two is a cake number. It is the maximum number of pieces into which a cube- or cylinder-shaped cake can be divided using just six cuts.

With one cut you can only get two pieces. With two cuts you can get four pieces. With three cuts, your first attempt might get you six pieces, while a little more thought shows that you can easily get seven. In fact, it is possible to get eight, by slicing horizontally through the middle of the cake before cutting into four from above. This doubling pattern (1, 2, 4, 8) then breaks down, since 15 is the answer for four cuts, and 26 for five cuts. The general problem is solved, with the solution for n cuts being $(n^3 + 5n + 6)/6$.

A far more difficult problem is to cut a cake in such a way that everyone believes they are getting their fair share...

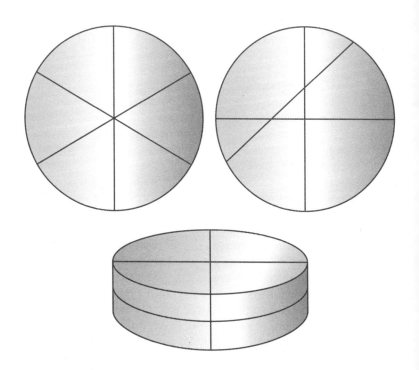

Dividing a cake into six, seven and eight pieces using just three cuts.
Can you fathom how to arrange six cuts to arrive at forty-two pieces?

43

In a fast-food restaurant, chicken nuggets are sold in boxes of six, nine or twenty. The question is: what is the largest number of nuggets that cannot be ordered with a combination of these boxes? For example, you can get twenty-one nuggets by ordering two of the six-nugget boxes and one nine-nugget box. But it is impossible to order exactly twenty-two nuggets.

This problem is an example of a more general puzzle called the coin problem: given a number of coins of different values, what is the largest amount of money that cannot be made? If the different coins have no common divisor (that is, no number that divides all of them exactly) then there will be a largest number, beyond which all values are possible. In the chicken nuggets problem, six, nine or twenty have no common divisor, and so it turns out that forty-three is the largest number of nuggets that cannot be ordered. Mathematicians have a formula for solving the coin problem where there are two types of coin, but there is no general answer for three or more.

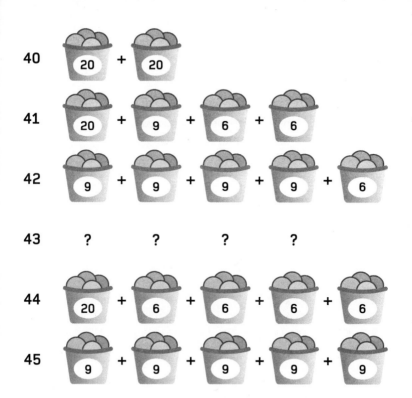

40	20 + 20
41	20 + 9 + 6 + 6
42	9 + 9 + 9 + 9 + 6
43	? ? ? ?
44	20 + 6 + 6 + 6 + 6
45	9 + 9 + 9 + 9 + 9

Forty-three is the largest number of nuggets that cannot be ordered from a combination of boxes containing six, nine and twenty nuggets.

44

Many people are familiar with Fibonacci numbers: the sequence 0, 1, 1, 2, 3, 5, 8, 13, … where each number is the sum of the previous two numbers (see page 50). Instead of being a Fibonacci number, forty-four is a tribonacci number. In this case, each number in the sequence is the sum of the previous three numbers. The traditional start for tribonacci numbers is 0, 0, 1, which gives the sequence as: 0, 0, 1, 1, 2, 4, 7, 13, 24, 44, …

Just as the ratio of adjacent terms in the Fibonacci sequence gets closer and closer to a special number called the golden ratio (see page 328), so too does the ratio of adjacent terms in the tribonacci sequence. In this case the number, called the tribonacci constant, is approximately 1.839. The tribonacci constant pops up unexpectedly in the geometry of the snub cube, one of the Archimedean solids, in describing the locations of its corners (see opposite). There are also tetranacci numbers, pentanacci numbers, hexanacci numbers, and so on, each having their own special constant.

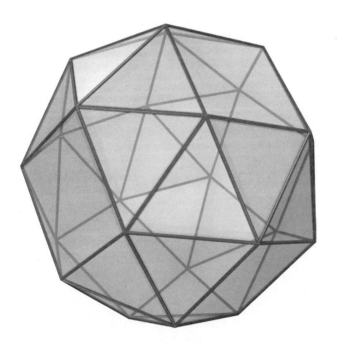

The snub cube is built from squares and equilateral triangles. Since all its faces are regular polygons, and all its vertices identical, it is an Archimedean solid. Its vertices have coordinates that are permutations of $(\pm 1, \pm 1/t, \pm t)$, where t is the tribonacci constant.

48

How many symmetries does a cube have? In other words, what are all the ways that a cube can be rotated or reflected to keep it looking exactly the same? The answer is forty-eight: can you find them all before you carry on reading?

First of all, you can rotate the cube by either 90°, 180° or 270° (a quarter turn, a half turn or a three-quarter turn) by holding the cube at the centres of opposite faces. Since there are three pairs of faces, this gives you 3 × 3 = 9 possible rotations. If you hold the cube by opposite corners, you can make 120° or 240° rotations (one-third and two-thirds of a turn). There are four pairs of corners, so this gives you 2 × 4 = 8 more rotations.

You can also make 180° rotations, by holding the cube at the centres of opposite edges – for example, the middle of the top back edge and the middle of the front bottom edge. With six pairs of edges, that's another six rotations. How did you do?

Symmetry in a cube

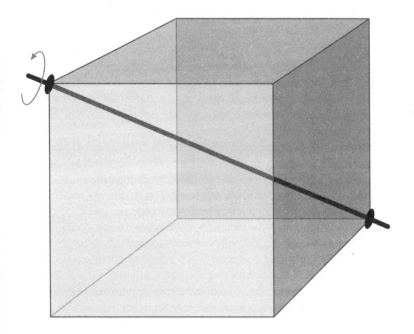

To rotate a cube by one-third of a turn to leave it looking the same, hold the cube by opposite corners.

49

A square number, forty-nine is equal to 7 × 7, which also means that it is the area of a square whose side lengths are each seven units long. It's the first square number whose digits (4 and 9) are themselves both square numbers.

Forty-nine is also equal to the sum of the first seven odd numbers: 49 = 1 + 3 + 5 + 7 + 9 + 11 + 13. It is no coincidence that forty-nine is both a sum of consecutive odd numbers and a square number. In fact, every square number is a sum of consecutive odd numbers. The easiest way to understand this is to look at the picture opposite.

A squareful number is one that can be divided exactly by a square number. For example, the number forty-eight is squareful because it can be divided by 16 = 4 × 4, and the number fifty is squareful because it can be divided by 25 = 5 × 5. It turns out that forty-nine is the smallest number with the property that itself and its two neighbours are all squareful.

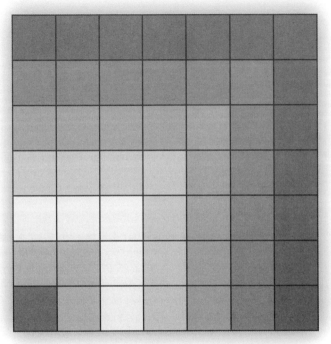

The odd numbers 1, 3, 5, 7, etc are drawn here as L-shapes radiating out from the bottom left tile. Each time you add an L-shape you get a square, so this picture shows you why the sum of the first *n* odd numbers is equal to *n* × *n*. For example, 1 + 3 + 5 (the first three L-shapes) create a 3 × 3 square, and the first seven odd numbers create a 7 × 7 square.

50

The phrase 'fifty-fifty' is common in the English language. It means something divided into two equal parts ('they split the cake fifty-fifty') and is often used in probability to mean that an event is as likely to happen as not ('there was a fifty-fifty chance of the cake being eaten'). The phrase comes from the fact that 50% is equal to one half ($\frac{1}{2}$), being equal to fifty out of one hundred.

Mathematically, fifty is interesting in that it is the smallest number expressible as the sum of two (non-zero) square numbers in two different ways. We can write 50 as $25 + 25 = 5^2 + 5^2$, and we can also write it as $1 + 49 = 1^2 + 7^2$. A result proved by Diophantus of Alexandria in the third century CE says that if you take two numbers that are each the sum of two squares, and multiply them together, the answer will be the sum of two squares in two different ways. This fact about fifty, therefore, follows from 50 being equal to 5 × 10, since 5 and 10 are the two smallest numbers that are the sums of two squares ($1^2 + 2^2$ and $1^2 + 3^2$).

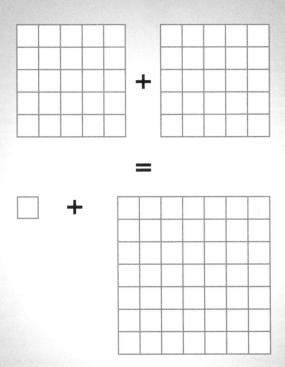

Fifty is the smallest number that is the
sum of two squares in two different ways.

57

Have you ever made the mistake of thinking that fifty-seven was a prime number? If so, you are in good company. Alexander Grothendieck, a brilliant twentieth-century geometer, was a notoriously abstract thinker. When asked to give a concrete example with a prime number, he said, 'Alright, take fifty-seven'. Since then, fifty-seven has been jokingly named the 'Grothendieck prime'. In fact, fifty-seven is a semiprime, being the product of two other primes, three and nineteen.

There is a whole category of numbers called 'April Fool primes'. Each is a composite number that ends in a 1, 3, 7 or 9, which makes them 'look' prime. For example, 87, 119 and 253 are April Fool primes.

Despite fifty-seven being a prime trickster, if someone shows you a number below 1000 ending in 57, your best guess is that it will be prime. This is because 57 is the most common two-digit ending among primes up to 1000.

57

Alexander Grothendieck

60

Being divisible by more numbers than any number below it, sixty is a highly composite number. It is also the smallest number to be divisible by all the numbers one to six.

A base-60 number system is called sexagesimal. This way of representing numbers was first used by the ancient Sumerians, who passed the system on to the Babylonians. It had symbols for one and ten, which were then used to make the numbers from one to fifty-nine. The placement of these symbols in columns determined how many powers of sixty they represented. The Babylonian use of base-60 is what gives us sixty seconds in a minute and sixty minutes in an hour, along with 360 degrees in a circle. Sexagesimal systems work particularly well to represent fractions, since the high divisibility of sixty means that none of the small fractions require an infinite number of symbols. For example, one-third of an hour is twenty minutes, and one-sixth of an hour is ten minutes. With one hundred minutes in an hour, these numbers would not be as neat.

Babylonian numbers

1	11	21	31	41	51
2	12	22	32	42	52
3	13	23	33	43	53
4	14	24	34	44	54
5	15	25	35	45	55
6	16	26	36	46	56
7	17	27	37	47	57
8	18	28	38	48	58
9	19	29	39	49	59
10	20	30	40	50	60

Babylonian numbers one to sixty using symbols for units and tens (see 1 and 10 above). There was no symbol for zero until later in history (see page 8).

62

A classic puzzle of recreational mathematics is the 'mutilated chessboard'. A standard 8 × 8 chessboard has one pair of opposite corners removed. You are given a pile of dominoes, with each domino large enough to cover two of the chessboard squares. Is it possible to place the dominoes so that every one of the remaining sixty-two squares on the chessboard is covered?

The puzzle is frustrating because it feels as if there should be an easy solution, yet no matter how you place the dominoes, you will keep getting stuck. In fact, the puzzle is impossible.

The explanation for this is surprisingly simple. Each domino will cover exactly one white and one black square on the board. Since there are thirty white squares but thirty-two black squares, the dominoes cannot possibly cover the entire board. It turns out that if two squares of opposite colour are removed, a domino tiling of the board is possible.

Mutilated chessboard

Can you cover this mutilated chessboard with dominoes,
where each domino covers two squares?

64

Is it possible for a number to be both a perfect square and a perfect cube? Yes, it is! The smallest example (apart from 1) is sixty-four, which is equal to 8 × 8 and 4 × 4 × 4. Such numbers can always be written as a sixth power x^6 (six instances of a number x multiplied together). Our number sixty-four is equal to 2^6, meaning we can write it as $(2^2)^3$ and $(2^3)^2$.

Robert Recorde was a Welsh physician and mathematician living in the sixteenth century. He is celebrated as the person who invented the equals sign (=) and who popularized the use of the addition sign (+). Recorde developed a great word for a sixth power: a 'zenzicube'. The word 'zenzic' is derived from the medieval Italian *censo*, meaning 'squared'.

Taking this even further, Recorde's word for an eighth power was a 'zenzizenzizenzic' – the square of a square of a square. This is currently the word containing the highest number of zs in the *Oxford English Dictionary*.

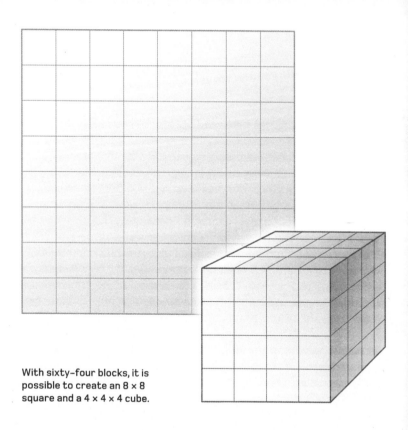

With sixty-four blocks, it is possible to create an 8 × 8 square and a 4 × 4 × 4 cube.

70

Seventy is 'weird' – a technical term that requires a few definitions. First, an abundant number is one whose proper divisors (see page 20) add up to more than itself. For example, twelve is abundant because its proper divisors (1, 2, 3, 4, 6) add up to sixteen. Second, a semiperfect number is one that is equal to the sum of some of its proper divisors. So twelve is semiperfect because it is equal to 6 + 3 + 2 + 1.

A weird number is one that is abundant but not semiperfect, and these are relatively rare. The smallest example is seventy: its proper divisors are 1, 2, 5, 7, 10, 14 and 35, which sum to seventy-four, but no combination of these numbers sums to seventy. The next weird number after seventy is 836.

We know that there are infinitely many weird numbers, but it is an open problem in mathematics whether any odd weird numbers exist. If there is such an example, it must be bigger than 10^{21}, or a thousand billion billion.

The smallest weird number

72

Tiling patterns are a familiar sight: you see them every time you walk on a pavement, gaze at a bathroom wall or look at the brickwork on a house. In regular tiling, you could trace the pattern onto paper and slide the paper so that the tracing lines up perfectly with a new set of tiles. This would be true no matter how large the area – you can assume an infinite wall and infinitely large piece of paper. Another way of saying this is that most tilings are periodic – they repeat themselves.

A Penrose tiling is aperiodic. If you trace its pattern onto paper, there is nowhere you can move the paper that would line up again with the tiling. Despite being made up of only two simple tiles, the pattern never repeats. The interior angles of these tiles feature one that is 72° and that is because the tiles are created from a pentagon. The exterior angles of a pentagon are also 72°. Penrose tilings can be constructed to have a five-fold symmetry – something that is impossible in regular tiling patterns.

A Penrose tiling

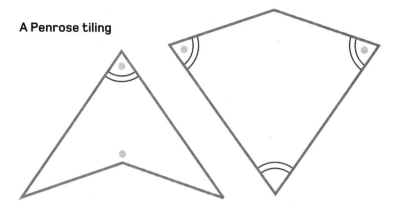

A Penrose tiling is made up of darts (left) and kites (right). A kite has three interior angles of 72° and one of 144° (2 × 72°). The two pointy angles in the darts are each 36° (half of 72°). When kites and darts are placed into a tiling, the dots in the corners must be placed next to each other to ensure an aperiodic tiling.

90

In geometry, an angle of 90° is one quarter-turn of a full circle. It is commonly known as a right angle, short for an 'upright' angle. This is because a vertical (upright) line placed perpendicular to a horizontal line creates two right angles. This description dates back to the ancient Greek geometer Euclid, who defined a right angle in this way. Euclid created a list of basic assumptions about geometry (called axioms) from which all other results could be proven. One of his five axioms was that all right angles are equal to each other – an important fact upon which all other geometry rests.

Right-angled triangles are the most important of all triangles. The trigonometric functions sine, cosine and tangent (sin, cos and tan) are all defined using the ratios of sides in a right-angled triangle. Any other triangle can be split into two right-angled triangles by drawing a line from a corner to the opposite side. Since all polygons can be split into triangles, all shapes are made of right-angled triangles.

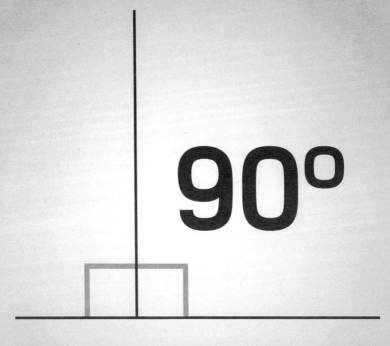

If you draw an upright line so that the angles on either side of it are equal, the two angles are called right angles. They are usually indicated by completing a small square in the angle they are in.

92

T he 'eight queens' puzzle asks how many ways we can place eight queens on a chessboard so that no queen is attacking any other. For non-chess fans, this means placing eight pieces on an 8 × 8 board so that no two pieces are in the same row, column or diagonal. It is not too difficult to find a way to place the queens, but it is tricky to find all solutions. Although the problem was posed back in 1848, it lends itself particularly well to computer programming, as a computer can systematically run through all possibilities. It turns out that the eight queens puzzle has ninety-two solutions, although some of these are rotations or reflections of one another. Discounting these repetitions gives twelve 'fundamental' solutions.

More generally, we can ask about the number of ways of placing n queens on an $n \times n$ board. This has been solved up to $n = 27$, although nobody has found a formula to predict the answers or describe the behaviour of the numbers. Interestingly, the 'five queens' puzzle has more solutions than the 'six queens' puzzle.

The eight queens puzzle

Here is one way you can place eight queens on a chessboard so that no queen attacks any other queen.

99

There is a very counterintuitive result in mathematics called the potato paradox. Suppose I have 100kg of potatoes that are 99% water by weight. I let them dehydrate until they are 98% water. How much do the potatoes weigh now? Your likely guess is 99kg or 98kg. After all, if 100kg was 99% water, then it should only be a small decrease to get 98% water. But the answer is, in fact, 50kg.

The first thing to notice in solving the problem is that the non-water pure potato weight remains fixed at 1kg both before and after the dehydration. Initially this weight is 1% of the total, but after dehydration it is 2% of the total. So, another way of phrasing the problem, which makes the final answer easier to grasp, is how much water needs to be removed so that the pure potato weight is worth twice as much of the total? The answer is 'about half'. Numerically we can reason as follows. If 1kg of weight is 2% of the total, then 0.5kg is 1%, and 50kg is 100%. So 50kg is the total weight after dehydration.

100

One hundred is a square number (equal to 10×10) and the sum of the first nine primes. It is actually the smallest square number that is a sum of first primes. According to the ancient Greek mathematician Nichomachus, the sum of the first n numbers squared is also the sum of the first n cubes. In this case, 100 is the sum of the first four cube numbers: $100 = 1^3 + 2^3 + 3^3 + 4^3$. This follows from the fact that $100 = 10^2 = (1 + 2 + 3 + 4)^2$.

The concept of percentages is based on the number 100 (*per cent* is Latin for 'per hundred'). A percentage is a ratio out of one hundred and is denoted by the % sign. For example, 50% means 50 out of 100, or a half; 100% means the whole amount.

One hundred also appears in metric systems of measurement and currency. There are one hundred pence in one pound sterling, and one hundred cents in one dollar or euro. In the Celsius temperature scale, one hundred degrees is the boiling point of water.

Nichomachus's theorem

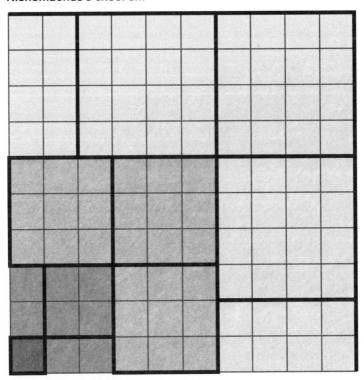

A graphical representation of why $(1 + 2 + 3 + 4)^2$ is equal to $1^3 + 2^3 + 3^3 + 4^3$.

101

This his number is a palindromic prime, meaning it is a prime number that reads the same forwards as backwards. All palindromic primes have an odd number of digits, since those with an even number of digits are always divisible by eleven.

To get the reciprocal of a number, you divide one by that number. The reciprocal of 101, that is $1/101 = 0.00990099\ldots$, repeats every four digits. It is the only prime with this property, making it a 'unique prime'. Any prime whose reciprocal has a unique period among primes is called unique. The first three unique primes are 3, 11 and 37. Unique primes are rare (there are only twenty-three among all primes with up to a hundred digits) but it is conjectured that there are infinitely many of them.

Add up the five consecutive primes, $13 + 17 + 19 + 23 + 29$, and you get 101. You also get 101 if you add together all the products $p \times q$ involving the first four primes 2, 3, 5 and 7. That is, $101 = (2 \times 3) + (2 \times 5) + (2 \times 7) + (3 \times 5) + (3 \times 7) + (5 \times 7)$.

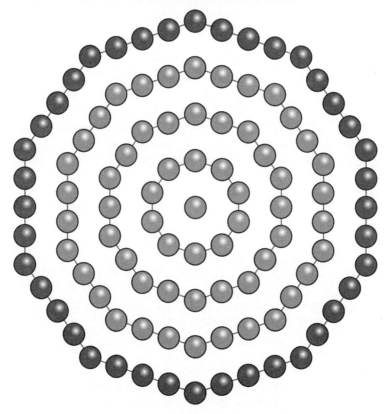

Graphically, 101 can be drawn as a centred decagonal number, with decagons drawn in increasing sizes and radiating out from a central dot.

108

Being equal to $1^1 \times 2^2 \times 3^3$, 108 is a composite number that is also a hyperfactorial. It can be written as the sum of a cube and a square in two different ways, being 27 + 81 ($3^3 + 9^2$) and 8 + 100 ($2^3 + 5^2$).

The interior angles of a regular pentagon are all 108°. One consequence of this is that regular pentagons cannot be used to tile the plane (see page 38), since 108 does not divide evenly into 360. Pentagons cannot even be combined with any other regular polygon to tile the plane (as, for example, octagons can be combined with squares). However, pentagons do tile a sphere: combining twelve pentagons generates a three-dimensional shape called the dodecahedron, which is one of the five Platonic solids (see page 15).

Should you ever need a 108° angle, you can easily fold a regular pentagon from a strip of paper by tying the paper into an overhand knot and pressing it flat (see opposite).

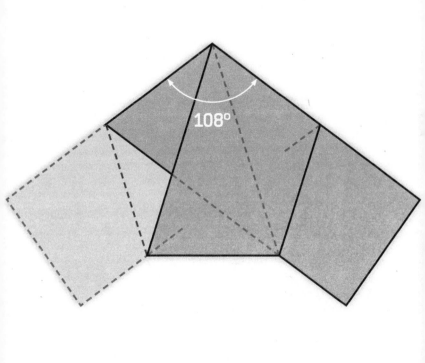

112

Squaring the circle – finding a square with the same area as a given circle – is a famously impossible problem, but what about squaring the square? It would be silly to ask for a square with the same area as another square, but the problem becomes interesting if we ask whether a square can be drawn as a collection of smaller squares.

To make the problem more difficult still, mathematicians insist on looking for solutions where all the smaller squares are of different sizes (a 'perfect' squared square) and where none of the small squares can be rearranged with each other to form any other rectangles or squares (a 'simple' squared square). The first perfect squared square was discovered in 1939, when German mathematician Roland Sprague found an example of fifty-five different squares fitting together in a big square with sides of length 4205. The simplest simple perfect squared square (with the smallest number of squares) has side lengths of 112 and contains twenty-one squares (see opposite).

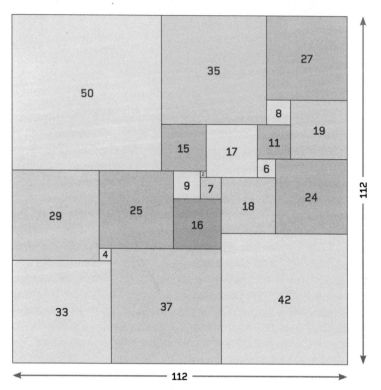

This 'square of squares' has size 112 × 112 and is made up of twenty-one smaller squares of different sizes. There are three simple perfect squared squares that are smaller (length 110) but they each contain twenty-two squares.

113

The Gauss circle problem is an unsolved problem in mathematics and involves counting points inside circles. With a circle of radius r centred at $(0,0)$ on a coordinate grid, the Gauss circle problem asks how many points inside or on the edge of the circle have integer (positive or negative whole number) coordinates. Such points are called lattice points. In algebraic terms, you want to find all the integers m and n so that $m^2 + n^2 \leq r^2$.

Think of each square on the coordinate grid as being associated with one lattice point. Each square also has an area of one. The area of the circle, πr^2, is therefore an approximation for the count of lattice points. When r is six, the two answers agree: there are 113 lattice points, and the area is 113 (rounded to the nearest whole number). However, for most other values of r there is a discrepancy between the two numbers, with sometimes the circle's area being bigger, and sometimes the lattice point count being bigger. Nobody has yet been able to quantify the discrepancy.

The Gauss circle problem

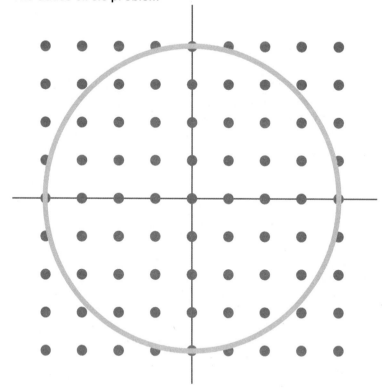

120

The number 120 is equal to five factorial, or 5 x 4 x 3 x 2 x 1, expressed as 5! (see page 56). It is a highly composite number, meaning it has more divisors than any smaller number. It is also one of only three known factorial numbers that are one less than a perfect square – it is unknown if more exist.

This number is the sum of the first four powers of three, $3^1 + 3^2 + 3^3 + 3^4$, as well as the sum of four consecutive powers of two, $2^3 + 2^4 + 2^5 + 2^6$, and four consecutive primes: 23 + 29 + 31 + 37.

Adding up the whole numbers from one to fifteen gives an answer of 120. This makes 120 a triangular number. Furthermore, if you add up the first eight triangular numbers, you also get an answer of 120, and this makes it a tetrahedral number. This means it is possible to stack 120 spheres in a triangular pyramid shape.

A football can be rotated or reflected in
120 different ways to leave it looking the same.

144

Equal to 12 × 12, 144 is a square number. It is also the twelfth Fibonacci number (see page 50). Apart from zero and one, 144 is the only Fibonacci number that is also a square. In fact, the numbers 0, 1, 8 and 144 are the only Fibonacci numbers that are perfect powers of other numbers.

Something of a rarity, 144 is one of the few sum-product numbers, which means it is equal to the sum of its digits (1 + 4 + 4 = 9) multiplied by the product of its digits (1 × 4 × 4 = 16). The only other such numbers are 0, 1 and 135.

Fifteen numbers divide exactly into 144. However, 144 does not qualify as a highly composite number, since there is a smaller number with even more divisors (120, with sixteen divisors). Still, the high divisibility of 144 makes it a useful number. The game Mahjong has 144 tiles, since this allows the tiles to split into thirty-six tiles of three different suits, plus sixteen wind tiles, twelve dragon tiles and four each of flowers and seasons.

144 is a dozen dozens,
or one gross.

x12

168

The Fano plane is a mathematical object made up of seven points and seven lines. Every line passes through three points, and every point lies on three lines, making this a wonderfully symmetric object. In fact, there are 168 ways of permuting the points so that the geometry is preserved.

You may also notice the following two things. If you pick any two lines, they meet at a unique point, and if you pick any two points there is a unique line joining them. These properties tell mathematicians that the Fano plane is an example of a projective space. Projective spaces were inspired by Renaissance artists, who were trying to draw accurate pictures of our three-dimensional world. Parallel lines never meet each other, but artists realized that if they drew two straight lines meeting at a point on the horizon (the point 'at infinity') then they could create the illusion of parallel lines and depth of field. Projective geometry remains important today in computer graphics, cryptography and quantum physics.

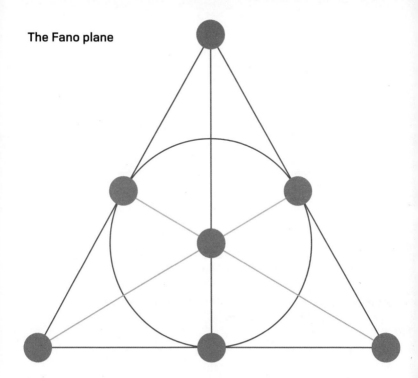

The Fano plane

In this picture, there are 168 ways of rearranging the points so that any three points that start out along a line are still on a line afterwards.

180

With more divisors (eighteen) than any smaller number, 180 is a highly composite number. It can be written as the sum of six consecutive primes (19 + 23 + 29 + 31 + 37 + 41) or as the sum of eight consecutive primes (11 + 13 + 17 + 19 + 23 + 29 + 31 + 37). It is also equal to the product of the squares of the first two primes, multiplied by the sum of the first two primes: $180 = (4 \times 9)(2 + 3) = 36 \times 5$.

Two angles that lie along a straight line will sum to 180°, as this is the angle of half a full circle (360°). The interior angles in any triangle sum to 180°, with the converse also true: any three positive numbers that sum to 180 can be the three angles in a triangle.

When a triangle is drawn on a sphere, their interior angles will sum to more than 180° (see page 152), and when triangles are drawn on a saddle-shaped surface, the angle sum will be less than 180°.

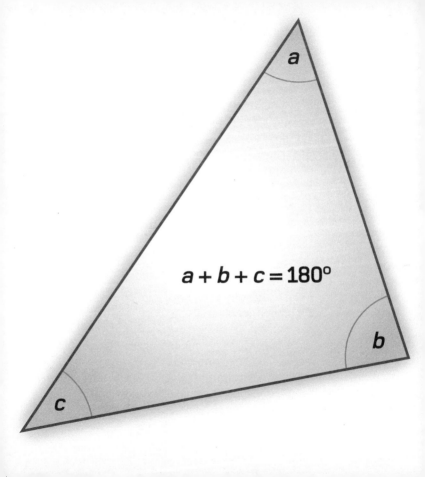

200

A primeable number is a composite number that can be turned into a prime by changing just one of its digits. For example, 125 can be made prime by changing the last digit to a seven, while 2863 can be made prime by changing the six to a four, giving the prime 2843. The same cannot be said for 200, however, which is the smallest unprimeable number. Changing either of the first two digits cannot make a prime, since the number will be a multiple of ten. Our only hope would be to change the last digit to an odd number, but 201, 203, 205, 207 and 209 are all composite. There are infinitely many unprimeable numbers, although the first unprimeable number ending in an odd number is not until 212159.

A sister concept relates to prime numbers, where a prime that cannot be turned into another prime by changing a single digit is known as weakly prime. For example, forty-one is not weakly prime, since it can be changed to the prime forty-seven. The smallest weakly prime number is 294001.

700 230 209 205 207

200

300 220 290

It is not possible to create a prime number by changing the number 200 by just one digit.

220

The proper divisors of 220 – numbers less than itself that divide into it exactly – are 1, 2, 4, 5, 10, 11, 20, 22, 44, 55 and 110. Add them all together and they make 284. In a very pleasing mathematical coincidence, the proper divisors of 284 add up to 220! Mathematicians call 220 and 284 amicable numbers. They have been known about since the Pythagoreans in the sixth century BCE, who considered them to have magical properties. Although more than one billion pairs of amicable numbers have been discovered, there are many unanswered questions about them. We do not know if there are infinitely many pairs, whether it is possible to have a pair where one number is odd and the other even, nor whether there is a pair that is relatively prime – that is, where the numbers share no common factor.

As well as there being amicable numbers, there are also social numbers. These are chains of numbers in which the proper divisors of one number add to the next, and so on in a circle. An example of this is 1264460, 1547860, 1727636 and 1305184.

230

The number 230 is composite and is a product of the three primes 2, 5 and 23. It is the first number where both it and the next number (231) are products of three primes.

There are 230 three-dimensional space groups. A space group represents a different combination of symmetries that a crystal can have. For this reason, they are also called crystallographic groups. Their analogues in two dimensions are the seventeen wallpaper groups (see page 42). A crystal is any arrangement of atoms in three dimensions such that the structure can be translated (moved) so that the atoms seem unmoved. As well as translation symmetry, crystals can exhibit rotation symmetries, reflections, screw symmetries (translation combined with rotation) and glide plane symmetries (translation combined with reflection).

Not every type of rotation is possible in a space group. For example, no crystal exhibits a five-fold symmetry.

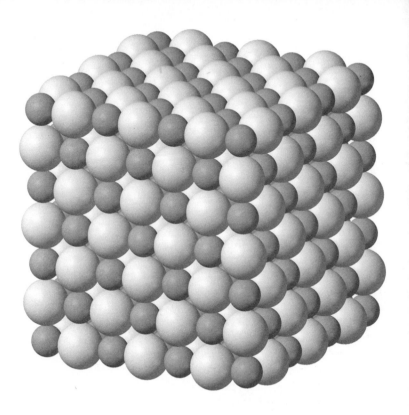

The atoms in salt (sodium chloride) form a crystal with a cubic lattice arrangement. Their symmetry is one of the 230 space groups.

248

Symmetry is fundamental to our grasp of the universe. In terms of physics, a symmetry is any transformation that doesn't change the laws of physics. For example, physics is the same no matter where in the universe we are (space symmetry), no matter when we exist (time symmetry) and no matter which direction we face (rotation symmetry). These examples are all continuous symmetries: we can smoothly rotate around any angle, or move steadily from one point to another. The objects in mathematics that capture the idea of continuous symmetries are called Lie groups. Most fall into one of several families. There are five exceptions to this, labelled E_6, E_7, E_8, G_2 and F_4. Of these, E_8 is the largest and most complicated. It is a 248-dimensional object, which means that every point inside it needs 248 numbers to describe its coordinates. While standard Lie groups have been used in physics to explain the Standard Model and conservation laws, E_8 has appeared in string theory and theories of super-symmetry. The hope is that it may someday provide a path to unify the theories of gravity and quantum mechanics.

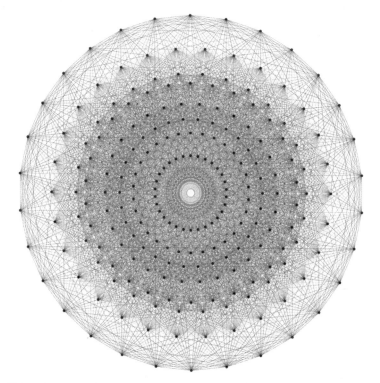

This picture provides a way of understanding the 248-dimensional shape that is E_8. It can be broken down as an eight-dimensional object plus a dimension for each of the 240 lines drawn in this diagram, which themselves each correspond to a vertex in a shape called the Gosset polytope.

249

What is a knot? It is a piece of string that follows a tangled path in three-dimensional space. Knots became part of mathematics in the nineteenth century due to the efforts of Scottish scientists Lord Kelvin and Peter Guthrie Tait, who believed (wrongly) that atoms could be described by the shape of knots. In a mathematical knot, the two free ends of the string are fused together to make a continuous loop. Two knots can then be considered 'the same' if one can be moved about in space until it looks like the other. They can be characterized by squashing them flat and counting the (minimum) number of times the string crosses over itself – called the crossing number. Tait set out to create tables of prime knots – those knots that cannot be tied as a collection of smaller knots on the same piece of string. He found that there was one knot with three crossings, one with four crossings, two with five, three with six and seven with seven. There are 249 prime knots with ten or fewer crossings, with the count getting into the millions for the knots with up to twenty crossings.

Prime knots

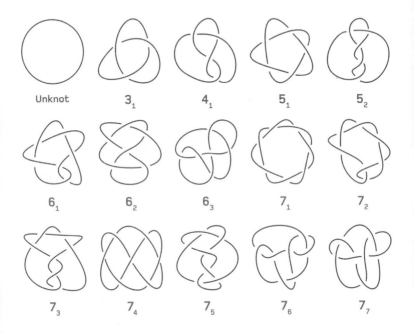

The first fifteen prime knots, characterized by their crossing numbers.
There is no general formula for the number of knots of a given crossing
number, and the problem of being able to tell knots apart is still one
being researched today.

256

The first number that is a square of a square of a square is 256. It is equal to 16^2 (16×16), but $16 = 4^2$ and $4 = 2^2$, meaning that 256 is $((2^2)^2)^2$. We can also write it as 2^8, or eight copies of two multiplied together.

This number is important in computing. A bit is a 'binary digit', the smallest unit of information in a computer. A bit can take two values, 0 or 1, and these are easily stored in a computer using the on/off of a switch. A collection of eight bits is called a byte, and this can take 2^8, or 256, different values. A byte is big enough to encode all basic characters of text (uppercase and lowercase letters, numbers, punctuation) as well as basic commands such as backspace, shift and enter. Just as each house in a street is the smallest unit of real estate to be given a separate address, so a byte is the smallest unit of memory that has its own address in a computer. This is why the capacity of your hard drive is measured in the number of bytes. (See also page 168.)

270

You might think I was crazy if I claimed to be able to draw a triangle containing three right angles. After all, you would tell me, the angles in a triangle always sum to 180°, while three right angles sum to 270°! But what is a triangle? It's a shape made up of three sides that are each straight lines. And what is a straight line? It is the line that joins two points across the shortest possible distance.

So, what happens if I draw my triangle onto a surface that is not flat – for example, onto the surface of the Earth? On Earth (which we assume to be a perfect sphere), the shortest distance between two points is the arc of a great circle – a circle whose centre is the same as the centre of the Earth. This explains why my triangle of 270° is a true triangle. In fact, any triangle I draw on a sphere will have angles that add up to more than 180° – the larger the triangle, the bigger the total. This type of geometry is essential for navigation, GPS systems, satellites and space travel.

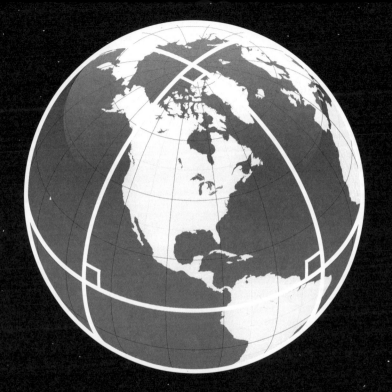

It is possible to draw a triangle on the Earth that starts at the North Pole, goes down to the equator, travels one-quarter of the way around the equator, then back up to the North Pole. Each angle in this triangle is 90°, so the sum of the angles is 270°.

360

A highly composite number, 360 has more divisors (twenty-four) than any smaller number. It is the smallest number divisible by all the integers from one to ten, except seven.

There are 360° in a circle. The reason for this may stem from there being approximately 360 days in a year, so the sun would advance in its path in the sky through one degree per day. Its high divisibility as a number makes 360 a better choice than the (more accurate) 365, since 360 can easily be divided into halves, thirds, quarters, tenths. It means we can divide the world into twenty-four time zones, each of which span 15° of longitude.

This number would also have made sense to the ancient Babylonians, as they used a base-60 (sexagesimal) system for counting (see page 106). Six equilateral triangles fit into a circle, each with an angle of 60°. In astronomy and geography, a degree is subdivided into sixty minutes, each of which is in turn divided into sixty seconds of arc.

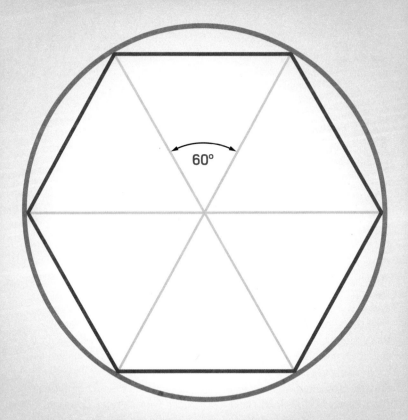

A full circle contains 360°, which can be divided
into six equilateral triangles of 60° each.

511

Prolific mathematician Paul Erdős (1913–96) published more papers than anyone else in history – over 1500 – mostly in collaboration with other people. This has led to the concept of an 'Erdős number'. A person has an Erdős number of one if they have co-authored a paper with Erdős. A person has an Erdős number of two if they have co-authored a paper with someone who has co-authored a paper with Erdős, and so on. There are 511 people with an Erdős number of one, and it is a matter of great pride for most professional mathematicians to have a number as small as possible. The author of this book has an Erdős number of five, the average among all mathematicians. Actors play a similar game, called Six Degrees of Kevin Bacon, in which they try to find the shortest sequence of films linking them to the actor. A person has a Bacon number of one if they have been in a film with him, of two if they have been in a film with someone who has been in a film with him, and so on. A small number of people in the world have a finite Erdős-Bacon number, which is the sum of the two numbers. These include Natalie Portman, Colin Firth and Kristen Stewart.

511

The Hungarian mathematician Paul Erdős spent most of his life living out of a suitcase and staying with other mathematicians. 'Another roof, another proof', was one of his many catchphrases.

561

The number 561 is composite: it is the product of three, eleven and seventeen. However, it is more notable for being the smallest Carmichael number, meaning that it looks prime through the lens of a common primality test called Fermat's Little Theorem. For this reason, Carmichael numbers are also called Fermat pseudoprimes.

To test if a number q is prime, mathematicians pick a number b and calculate $b^q - b$. If this is not a multiple of q, then q is not prime. If it is, then they try another value for b. The more values for b that pass the test, the surer mathematicians can be that q is prime. Carmichael numbers are problematic since they pass the test for every value of b even though they are composite.

Probabilistic primality testing like this is common in cryptography, where it can take too long to test true primality for large numbers. There are many different tests, but each will face the problem of pseudoprimes.

Fermat pseudoprimes

561	(3, 11, 17)
1105	(5, 13, 17)
1729	(7, 13, 19)
2465	(5, 17, 29)
2821	(7, 13, 31)

The first five Carmichael numbers and their factors.

600

The 600-cell is one of six highly symmetric four-dimensional shapes, called convex regular polytopes. It is built by gluing together 600 tetrahedra (triangular-based pyramids), with twenty meeting at each vertex and five meeting at each edge.

This shape is considered to be the four-dimensional analogue of the icosahedron, one of the five Platonic solids (see page 15), which is built from twenty equilateral triangles, with five triangles around each vertex. Just as we can visualize a three-dimensional cube by drawing a projection of it onto two-dimensional paper, so we can visualize the 600-cell by projecting it into three-dimensional space. The dual of the 600-cell is the 120-cell, which is composed of 120 dodecahedra joined together with four around each vertex. The 120-cell has 600 vertices, just as the 600-cell has 120 vertices. The 600-cell does not have any geometric analogues in higher dimensions. In fact, it is in four dimensions that there are the highest number of symmetric shapes: in all higher dimensions there are only three.

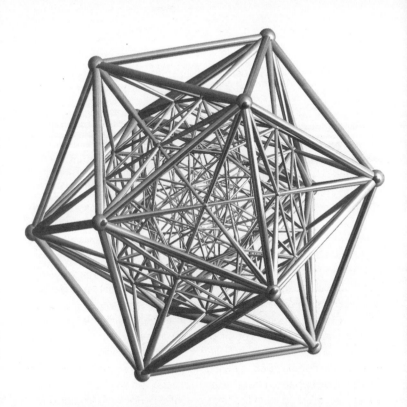

The 600-cell projected into three-dimensional space.
Every edge in this model would be the same size in four dimensions.

666

The ancient Romans represented numbers using an additive system of symbols. We still use this number system in many places today, including on clock faces and the names of monarchs – for example George VI. The basic idea is to create symbols for a set of numbers and then represent other numbers by adding those together. The Romans chose to make symbols for 1 (I), 5 (V), 10 (X), 50 (L), 100 (C), 500 (D) and 1000 (M). To make the number 2019, the Romans would write MXVIIII, which is 1000 + 10 + 5 + 1 + 1 + 1 + 1. Over time it became common to use IV and IX in place of IIII and VIIII respectively, which is called subtractive notation. The number 666 can be written in Roman numerals using exactly one of each of the symbols below 1000: DCLXVI. This number is interesting mathematically, being the sum of the squares of the first seven primes.

Fear of the number 666 is called hexakosioihexekontahexaphobia. The number appears in Christian mysticism as the 'number of the beast', as referred to in translations of the Book of Revelation.

871

Think of a whole number. If it is even, divide it by two. If it is odd, multiply it by three and add one. Repeat these instructions with your new number. For example, starting with seven, you get the following sequence of numbers: 7, 22, 11, 34, 17, 52, 26, 13, 40, 20, 10, 5, 16, 8, 4, 2, 1. The Collatz conjecture is an unsolved mathematical problem claiming that this process always returns to the number one, no matter what the starting number is. It has been tested up to numbers with twenty digits and, so far, no counterexample has been found, yet there is still no proof that it will always work. Out of all the numbers less than one thousand, 871 is the number that takes the most steps to reach one (178).

This conjecture has been around since 1937 and remains a frustration to mathematicians. There are two ways it could fail: a number could produce a trajectory that kept going for an infinitely long time, or its trajectory could end up going round and round in a cycle of the same numbers. Nobody can yet rule out these two options.

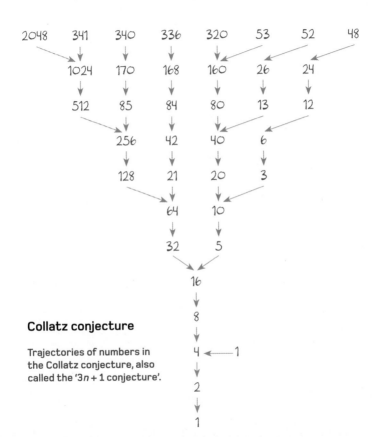

Collatz conjecture

Trajectories of numbers in the Collatz conjecture, also called the '3*n* + 1 conjecture'.

999

As well as being the emergency services telephone number in many countries, 999 is a mathematically interesting number. It is the largest three-digit integer and is the smallest multiple of twenty-seven whose digits also sum to twenty-seven.

This number is also the sum of three primes in (at least) two interesting ways. First, you can write 999 as 149 + 263 + 578. Notice that, between them, these three primes contain all the digits 1 to 9 exactly once. This is called a pandigital sum, which makes 999 the smallest pandigital sum of three-digit primes.

Second, you can write 999 as 271 + 331 + 397. These three numbers are called 'cuban primes' – not owing to any association with Cuba, but because they are formed from cube numbers. A cuban prime is the difference between two successive cube numbers: for example, $271 = 10^3 - 9^3 = 1000 - 729$. So 999 is the sum of three consecutive cuban primes.

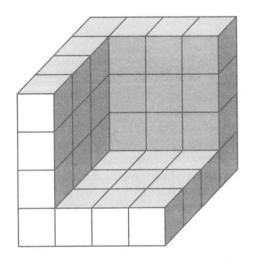

A cuban prime is a prime number that is calculated as the difference between successive cube numbers. For example, in this picture a 4 × 4 × 4 cube has had a 3 × 3 × 3 cube removed, leaving 64 − 27 = 37 cubes, so 37 is a cuban prime.

1000

One thousand is the first four-digit number in our decimal system. In Roman numerals 1000 is written M, representing the Latin *mille*, from which the words millennium (one thousand years) and millipede (creature with one thousand legs) derive. The Greek word for one thousand, *khílioi*, gives us our prefix kilo-. Thus, a kilometre is one thousand metres, a kilogram is one thousand grams and a kilowatt is one thousand watts. Yet beware, because a kilobyte often means 1024 bytes! Why is this? The number 1024 is equal to ten copies of the number two multiplied together, denoted 2^{10}, or 'two to the power of ten'. Counting in powers of two makes sense in computer science, since computers perform calculations using a base-2 system called binary. The difference between a metric kilobyte and a binary kilobyte may not be so much, but the discrepancy grows when we start to discuss megabytes, gigabytes and terabytes. To avoid confusion, metric kilobytes use a lowercase symbol (kB means 1000 bytes) and binary kilobytes use an uppercase symbol (KB means 1024 bytes).

Despite its common name, the millipede rarely has more than 200 legs,
although some species have as many as 650.

1089

Think of any three-digit number where the first and last digits differ by at least two – for example, 745. Reverse the digits (547). Subtract this number from the number you first thought of (745 – 547 = 198). Reverse the digits again (891) and add it to the last number (198 + 891). The answer will always be 1089. This surprising result is used as the basis for a number of magic tricks.

There is a hidden symmetry in the number 1089, which becomes apparent if you write it as nine hundreds, eighteen tens and nine units. Explaining why the trick works uses this fact plus some algebra about digit reversals. More patterns emerge when you look at the first nine multiples of this number. As you multiply by 1, 2, 3, 4, . . ., the first two digits of the answer go up by one each time, and the last two digits go down by one each time. So, $2 \times 1089 = 2178$, then $3 \times 1089 = 3267$. . . This pattern also means that when two numbers add to ten, their respective multiples of 1089 are reverses of each other. For example, $7 \times 1089 = 7623$, which is the reverse of 3×1089, since 3 and 7 add to 10.

$1 \times 1089 = 1089 \leftrightarrow 9 \times 1089 = 9801$

$2 \times 1089 = 2178 \leftrightarrow 8 \times 1089 = 8712$

$3 \times 1089 = 3267 \leftrightarrow 7 \times 1089 = 7623$

$4 \times 1089 = 4356 \leftrightarrow 6 \times 1089 = 6534$

$5 \times 1089 = 5445 \leftrightarrow 5 \times 1089 = 5445$

Multiplying 1089 by the numbers one to nine.

1260

A vampire number is a composite number with an even number of digits, with the property that it can be factorized into two numbers, each with half as many digits as the original, and so that these two numbers between them contain all the digits of the original (in any order). The first such number is 1260. This satisfies the definition since 1260 = 21 x 60. The two numbers in the factorization are called the fangs of the vampire number. There is an additional constraint that the two fangs cannot both have trailing zeros. This is to prevent uninteresting cases such as 126000, which could be written as 210 x 600.

Some vampire numbers have more than one pair of fangs. For example, 125460 can be written as 204 x 615 or as 246 x 510. There are also double vampire numbers, whose fangs are themselves vampire numbers. One example of this is 1047527295416280, which factors into the vampire numbers 25198740 and 41570622.

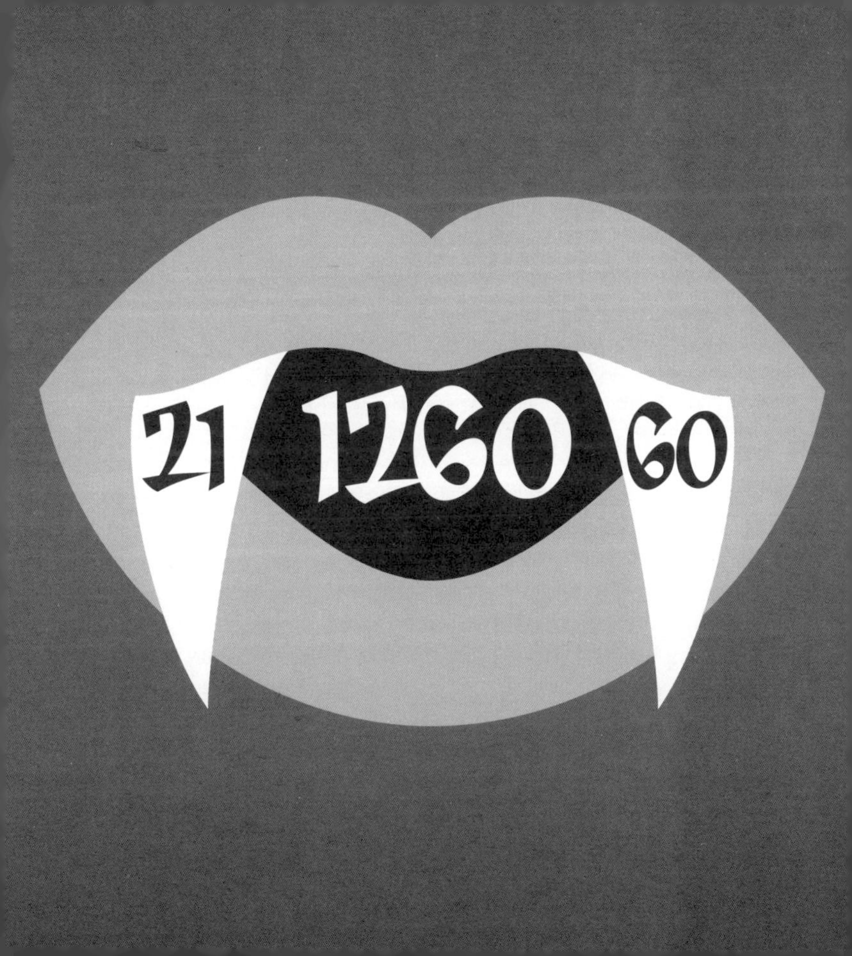
21
1260
60

1729

The number 1729 is part of a famous story involving the British mathematician G.H. Hardy and the Indian mathematician Srinivasa Ramanujan. Hardy went to visit Ramanujan in hospital in London, and on arrival remarked that he had travelled in a taxi with number 1729. Hardy lamented that the number was a dull one, whereupon Ramanujan replied that it was actually very interesting, being the smallest number that could be expressed as the sum of two (positive) cubes in two different ways. Indeed, 1729 can be written as $1^3 + 12^3$ and also as $9^3 + 10^3$.

Mathematicians have since generalized this concept, saying that the nth taxicab number is the smallest number that can be written as the sum of two positive cubes in n different ways. Thus 1729 is the second taxicab number. The first taxicab number is two, written $1^3 + 1^3$. The third taxicab number is 87539319, since it can be written as $167^3 + 436^3$, or $228^3 + 423^3$, or $255^3 + 414^3$. Taxicab numbers are only known up to $n = 6$.

G.H. Hardy (left) and Srinivasa Ramanujan (right). The nth taxicab number is also called the nth Hardy–Ramanujan number.

1936

In 1852, Francis Guthrie was colouring in a map of the counties of England. He noticed that he needed just four colours so that regions sharing a common boundary could always be given different colours. He wondered whether the same was true of all maps – a conjecture that became known as the four-colour conjecture. The conjecture didn't just apply to maps, but to any flat drawing divided into different regions. It is easy to find a map that cannot be coloured using three colours, but it was unclear whether a map existed that was complex enough to need five colours. The problem remained unsolved for over a hundred years. Then, in 1976, Kenneth Appel and Wolfgang Haken announced a highly controversial proof. Instead of creating an elegant argument for four colours always working, they had written a computer program to colour 1936 special maps, which they showed solved the problem. Today their proof is accepted by the mathematical community, and their 1936 maps have been reduced to 633. Nobody has yet found a non-computer-aided proof.

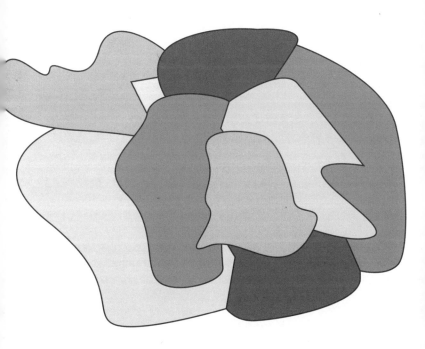

This map cannot be coloured using three colours so that adjacent regions are different colours – it needs at least four colours. But are four colours sufficient for every map?

2047

Mersenne numbers are those of the form $2^n - 1$. That is, they are one less than a power of two. Of particular interest are Mersenne numbers that are prime (see page 186). A Mersenne number can only be prime if the exponent n is prime. However, it is not true that if n is prime then the corresponding Mersenne number M_n will be prime. For example, the first prime p for which M_p is not prime is $p = 11$, giving $M_p = 2^{11} - 1 = 2047$, which is equal to 23×89.

Marin Mersenne, the French friar who first studied these numbers, knew that M_{11} was not prime. He attempted to write down a list of exponents for which M_p was prime, coming up with 2, 3, 5, 7, 13, 17, 19, 31, 67, 127, 257. Unfortunately this list has two types of error. It lists two values for which M_p is not actually prime (67, 257) and omits values for which M_p is prime (61, 89 and 107).

Today the search continues for Mersenne primes, with the eight largest known primes being of this type.

Marin Mersenne

3435

A Munchausen number is one that is equal to the sum of its digits each raised to the power of themselves. Apart from the number one, 3435 is the only Munchausen number, because it equals $3^3 + 4^4 + 3^3 + 5^5$. Controversially, 0 and 438579088 are Munchausen numbers if the convention is used that $0^0 = 0$. The name 'Munchausen number' was coined in 2009 by Daan van Berkel, who was inspired by the story of Baron Munchausen saving himself from drowning by 'raising himself up' by his ponytail.

A sister concept of narcissistic numbers requires each digit to be raised to the same power, which is the number of digits in the number. For example, 153 is narcissistic, since $153 = 1^3 + 5^3 + 3^3$. Narcissistic numbers are also finite in quantity, though there are eighty-eight of them compared with the two Munchausen numbers. Narcissistic and Munchausen numbers appear in every base (for example, in binary, ternary, and so on) but there are always only finitely many of them.

Baron Munchausen

5050

The story goes that when the great mathematician Carl Friedrich Gauss was ten years old, his schoolmaster set the class a challenge that was meant to keep the children occupied for some time. The task was to add up the numbers from one to one hundred. But young Gauss wrote down the answer within a minute: 5050. The teacher suspected him of cheating, but in fact Gauss had simply calculated the answer in a very clever way.

One method Gauss may have used is as follows: write out the numbers from 1 to 100 twice, one list above the other. In the top row, write the numbers going from 1 to 100 and in the bottom row write them from 100 down to 1. Adding up each column of numbers gives 101, and there are 100 columns, so the total is 10100. This must be double the answer Gauss's teacher was looking for, which is then 5050.

In general, the sum of the numbers from 1 to n is $n(n+1)/2$. Numbers of this form are called triangular numbers.

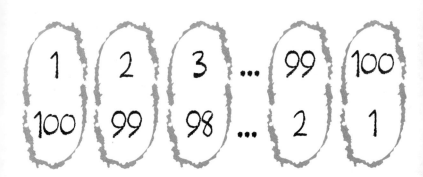

A speedy method for finding the totals
of the numbers from 1 to 100.

6300

The twentieth-century Dutch painter Piet Mondrian was famous for his abstract art, in which he divided rectangular forms into smaller rectangular shapes. His art has inspired a genre of mathematical puzzles called 'Mondrian art puzzles', which contain an intriguing unsolved problem. The 10 × 10 grid pictured is divided into a number of smaller rectangles, each of different dimensions. The score for a given dissection is calculated as the difference between the areas of the largest and the smallest rectangles. The challenge is to find the smallest solution. There seems to be no general formula for the answer, nor is it known whether an answer of 0 is ever possible.

Tantalizingly, there is a dissection of a square into different rectangles of the same size. It is called Blanche's dissection and is a square of side length 210 divided into seven rectangles, each of area 6300. However, these rectangles have sides of irrational lengths, so it does not solve the Mondrian art puzzle, where side lengths must be whole numbers.

Mondrian art puzzle

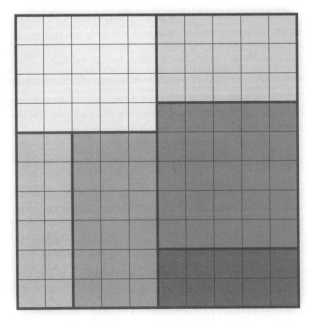

Our score for the dissection of this 10 × 10 grid is the difference
between the areas of the largest and smallest rectangles:
25 − 10 = 15. However, it is possible to do much better:
the lowest possible score is eight. Can you find the solution?

8191

A Mersenne prime is a prime number that is one less than a power of two. Such numbers, M_p, can always be written as $2^p - 1$, where p is another prime. The fifth Mersenne prime is 8191, which is $M_{13} = 2^{13} - 1$.

The difficulty with Mersenne primes is that not every prime p gives a Mersenne prime M_p (see page 178). For example, M_{11} is not prime even though 11 is prime. It might be reasonable to guess that if we start with a Mersenne prime M_p then M_{M_p} might be prime. This is true when p is 2, 3, 5 and 7, but sadly the conjecture fails with M_{13} since M_{8191} is not prime.

Numbers of the form M_{M_p}, where p is prime, are called double Mersenne numbers. None of the primes past $p = 13$ have provided any double Mersenne primes, with $p = 61$ being the next value yet to be tested. Being a number with nearly 700000 trillion digits, it is far too big for any current primality test.

$$M_{13} = 2^{13} - 1$$

26861

Think of any prime number other than two. If you divide your prime by four, the remainder will be either one or three. Mathematicians call these the $4n + 1$ primes and the $4n + 3$ primes respectively.

There are infinitely many of these two types of prime. Mathematicians have also shown that primes are distributed evenly between the two categories, so that half the primes are $4n + 1$ and half are $4n + 3$. However, if you keep a running total of how many primes are in each category, the $4n + 3$ primes are 'ahead' in the race almost all the time. The first time that the $4n + 1$ primes get ahead is after 26861. Their lead is short-lived, however: 26863 is a $4n + 3$ prime, and so the other side regain the lead and don't lose it again until 616841. Proved by J.E. Littlewood, the two sides switch places infinitely many times in the race. But understanding exactly how often the $4n + 3$ primes are in the lead is currently only possible when accepting the generalized Riemann hypothesis (see page 378).

53169

The On-Line Encyclopedia of Integer Sequences (OEIS) is a database of whole-number sequences. It was started in 1965 by Neil Sloane and now contains over 300,000 sequences.

Some sequences in the OEIS refer to the database itself, in a self-referential way. Sequence A53169 is one such example. A number n is in the sequence if, and only if, it is not in the sequence A$_n$. For example, sequence A000001 contains the number one, so one is not in A53169. Sequence A00004 does not contain the number four, so four is in A53169. But this generates a paradox: is 53169 in the sequence A53169? If it is in the sequence then it should not be, while if it is not in the sequence then it should be. This is a form of Russell's paradox, which was formulated in 1901 by Bertrand Russell to highlight contradictions in the new area of set theory. Sequence A53873 is also problematic: it contains all numbers n which are members of A$_n$. It is therefore impossible to say whether 53873 is a member of A53873, since either case is possible.

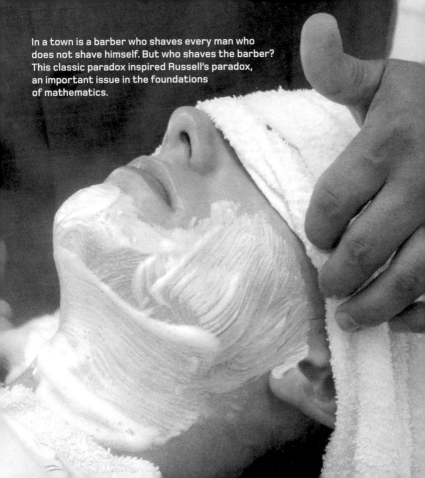

In a town is a barber who shaves every man who does not shave himself. But who shaves the barber? This classic paradox inspired Russell's paradox, an important issue in the foundations of mathematics.

65536

The smallest number with exactly seventeen divisors, 65536 is equal to 2^{16}, or two multiplied by itself sixteen times. Keen-eyed readers may notice that not only is 65536 a power of two, 2^{16}, but that sixteen is also a power of two: 2^4. Then four is also a power of two: 2^2. This means that 65536 is a 'power tower' of twos, written as shown opposite.

This operation of compound exponentiation is called tetration, and (in the case of 65536) it can be denoted by $^{4}2$. An alternative notation is Knuth's up-arrow notation, in which 65536 would be written as 2↑16, or 2↑2↑2↑2, or 2↑↑4, or 2↑↑↑3.

65536 is the only known power of two that does not contain another power of two in its digits. In particular, it contains no 1s, 2s, 4s or 8s. All powers of two up to 2^{31000} have been checked and no other examples have been found, although it is unknown whether another might exist.

$$2^{2^{2^{2}}}$$

65537

A prime number that is one more than a power of two, 65537 is a Fermat prime. All primes of this type are presented as shown opposite and 65537 is the case where $n = 4$. Pierre de Fermat conjectured that there were infinitely many Fermat primes. However, 65537 is the largest one anybody has found.

Since 65537 is a Fermat prime, a polygon with 65537 sides is constructible using a straightedge, or unmarked ruler, and a compass. This is true for all Fermat primes and has been known for a long time, but was only demonstrated in 1894 by Johann Gustav Hermes. He had spent ten years and 200 pages figuring out how to do it.

In RSA (Rivest–Shamir–Adleman) cryptography, 65537 is widely used because being a Fermat prime makes it simple to write and use in binary computations. Smaller Fermat primes, for example three, are also commonly used, but additional security comes from 65537 being so large.

$$2^{2^n}+1$$

78498

Predicting where the next prime number will be is a considerable challenge. A simpler task is to understand roughly how the primes are distributed among all numbers. The prime number count, written as $\pi(x)$, is the count of how many primes there are below a number x. For example, $\pi(10)$ is equal to 4, since there are four primes (2, 3, 5 and 7) below 10. How quickly does the prime count grow? If we double the number x, does the prime count $\pi(x)$ double too? The first person who answered this question was Carl Friedrich Gauss, in 1793. Before he was even sixteen years old, he had come up with a formula that was a good approximation to the count. For example, there are 78,498 primes below one million, and Gauss's formula predicts 72,382. Today it is called the prime number theorem, and says that $\pi(x)$ is approximately $x/\ln(x)$ (where $\ln(x)$ is the natural logarithm of x – see page 296). In plain English, the theorem says that a number with twice as many digits in it is half as likely to be prime. It shows that the primes become scarcer as you go up the number line.

Gauss's prime number theorem

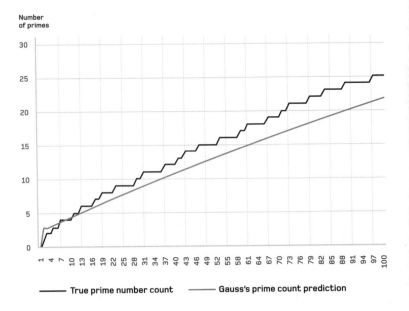

Number of primes

True prime number count   Gauss's prime count prediction

A graph showing the count of the number of primes up to 100 (in black) compared with Gauss's formula $x/\ln(x)$ (in grey). Gauss's formula captures the general shape of the prime curve, but the gap between the curves gets bigger as the numbers get larger. Quantifying this gap, and finding better formulae to reduce it, have led to the Riemann hypothesis (see page 378).

85900

Picture this scenario: you've gone into the big city to run some errands but you're short of time. You need to visit the post office, return a book to the library, pick up your dry cleaning, cash a cheque and buy some avocados. How do you figure out the shortest route that will take you where you need to go and back home again?

Companies such as Amazon and UPS solve problems like this every day on a massive scale, delivering thousands of parcels over a wide area. The right algorithm could save millions of dollars. This problem, called the Travelling Salesman Problem, is known among mathematicians as an example of an NP-complete problem. It is fast to check whether a solution is the best one yet, but slow to find a solution in the first place. It remains an open question with a million-dollar prize, as to whether a fast solution exists. The largest Travelling Salesman Problem to be solved exactly has 85,900 cities, and it took over 130 computing years to solve in 2006.

If you tried to work out the shortest route between these thirty cities by finding every possible route and then picking the shortest one, you would be waiting longer than the age of the universe to get the answer, even using the fastest supercomputer.

111221

****

Here is a fiendish number puzzle: find the next number in the pictured sequence. The puzzle is frustrating because the answer has nothing to do with the mathematics of these numbers. It is called the 'look and say' sequence because each number describes the previous one in the sequence: 1 (one); 11 (one 1); 21 (two 1s); 1211 (one 2, one 1); 111221 (one 1, one 2, two 1s); 312211 (three 1s, two 2s, one 1).

John Conway invented the sequence in 1986. Despite it not being a very mathematical puzzle, mathematicians have analysed these numbers in depth. For example, they have shown that the sequence will never contain any digits other than 1, 2 and 3. Conway discovered that the sequence can eventually be split into two strings of numbers written side by side, whose descendants never interfere with each other. Moreover, this phenomenon happens no matter the starting number in the sequence (except for the boring '22'), and there are 92 'elements' into which all sequences eventually 'decay'.

1, 11, 21, 1211, 111221, 312211, …?

142857

The number 142857 is an example of a cyclic number, where successive multiples give cyclic permutations of its digits. In a cyclic permutation, digits move from the front of a number to the back. This is the case when multiplying 142857 by 2, 3, 4, 5 and 6. For example, multiplying by 2 gives 285714 and multiplying by 3 gives 428571. It is no coincidence that 142857 is also the recurring decimal of $\frac{1}{7}$ (that is, these six digits repeat ad infinitum). If we take any prime number p whose reciprocal $\frac{1}{p}$ has a decimal that repeats after every $(p-1)$ digits (and no fewer), then those $(p-1)$ digits will be cyclic. Two more examples of this are seventeen and nineteen. However, these cyclic numbers begin with a zero. It turns out that 142857 is the only cyclic number that does not begin with a zero (excluding single digits and repeated digits such as 111).

Multiplying 142857 by 7 gives 999999, which makes sense if you remember that $\frac{7}{7} = 0.999999\ldots$ More surprisingly, perhaps, is that $142 + 857 = 999$ and $14 + 28 + 57 = 99$.

$$142857 \times 1 = 142857$$
$$142857 \times 2 = 285714$$
$$142857 \times 3 = 428571$$
$$142857 \times 4 = 571428$$
$$142857 \times 5 = 714285$$
$$142857 \times 6 = 857142$$

196560

Sending data to Earth from outer space requires clever mathematics to ensure that it is not lost or distorted. In the case of spacecraft, this involves something called a 'check digit'. The last number on your bank card is a check digit – the answer to a specific calculation involving the other numbers on the card. When you enter the number on a website, it is the check digit that ensures you haven't typed a mistake. The Voyager probes use a far more sophisticated set of check digits that not only detect errors, but correct them, too. The system is called binary Golay code. Within each twenty-four-letter word, the code can correct up to three different errors and detect a fourth.

The mathematics here involves a highly symmetric geometric structure called the Leech lattice. Just as a hexagonal lattice is the most efficient way to carve up two-dimensional space (see page 358), so the Leech lattice is the best way to partition twenty-four-dimensional space. Each ball placed in a hexagonal lattice has six neighbours; each ball in the Leech lattice has 196560!

NASA drawing of the Voyager spacecraft, currently hurtling away from Earth in interstellar space. Golay code exploits the efficiency of the highly symmetric Leech lattice to check Voyager's data as quickly as possible.

262144

The number 262144 is equal to 2^{18}, or two multiplied by itself eighteen times. It is interesting because it is also the fourth exponential factorial, written as shown opposite. Compound exponentiation is evaluated from right to left, so that you first compute $2^1 = 2$, then $3^2 = 9$, then $4^9 = 262144$.

Exponential factorials are denoted by $n\$$, so $262144 = 4\$$. The first three exponential factorials are $1\$ = 1$, $2\$ = 2$ and $3\$ = 9$. Regular factorials, denoted by $n!$ (see page 56), work via multiplication rather than exponentiation. So $4! = 4 \times 3 \times 2 \times 1$. Exponential factorials grow far more quickly than regular factorials: $5!$ is 120, but $5\$$ is a number with over 180000 digits.

262144 is also mathematically interesting as an example of a superperfect number: its divisors add up to 524287, whose divisors are 1 and 524287, which sum to $524288 = 2 \times 262144$. We know that all even superperfect numbers must be powers of two, but it is unknown whether any odd superperfect numbers exist.

1000000 (one million)

Derived from the Latin *mille*, meaning 'thousand', one million means a thousand thousand. The ancient Romans actually had no word for a number bigger than one hundred thousand, so a million would have had to be written as 'ten hundred thousand'. The ancient Greeks had no word for a number bigger than ten thousand.

The ancient Egyptians, however, who used an additive form of counting, had symbols for the powers of ten up to one million. The symbol for one million was that of the god Heh, who was the personification of infinity in ancient Egyptian mythology. Just as the concepts of 'million' and 'infinity' were conflated in ancient Egypt, so too do we often use the word 'million' in English to mean a very large number. For example, in expressions such as 'a million miles apart', or 'one in a million'. Billion and million are often confused, too, despite the fact that a billion is one thousand times bigger. A million seconds is 11.6 days, whereas a billion seconds last longer than 31 years.

1234321

Anyone who has daydreamed in a maths class or played with a calculator is likely to have found the following pattern. If you square the number eleven – that is, multiply it by itself – the answer is 121. If you square 111, the answer is 12321. If you square 1111, you get 1234321, and so the pattern continues, all the way up to 111111111. So, why does this pattern occur?

Consider the square of 1111 as an example. You can rewrite 1111 × 1111 as 1111 × (1000 + 100 + 10 + 1), and multiply this out to give 1111000 + 111100 + 11110 + 1111 = 1234321. Since 1111 has four digits and is shifted to the right four times, this gives us the rising and falling structure of the digits of the answer, with the biggest one being four.

These squares are all examples of palindromic numbers, which read the same forwards and backwards. There are infinitely many such numbers where both the square and the root of the number are palindromic.

11
121
12321
1234321
123454321
12345654321
1234567654321
12345678765431
12345678987654321

Squares up
to 111111111

3628800

The number 3628800 is equal to $1 \times 2 \times 3 \ldots \times 10$. In other words, it is the product of all the whole numbers up to ten. The mathematical notation for this is 10!, read aloud as 'ten factorial'. Ten factorial can, unusually, also be written as $1! \times 3! \times 5! \times 7!$, making it the largest factorial to be the product of (at least three) other factorials in an arithmetic sequence.

Factorials appear most commonly when counting permutations. For example, if I have ten pigs, there are 10! ways that I can order them. This is because there are ten choices for the first pig, then nine for the second pig, eight for the third, and so on. Factorials also appear in calculations for games, such as lotteries, where they can be quite counterintuitive. For example, what are your chances of winning a lottery in which you need to select the correct three numbers out of ten? You might imagine it is about 33%, but in reality it is closer to 0.8%, since there are $3! \times 7!$ permutations out of 10! that have the first three numbers correct, which is 1 in 120.

66600049

T think of a prime number. Does it have any digits that you
can cross out to make another prime number? For example,
with the prime 2243, I could cross out the middle two numbers
to leave 23, which is prime. Or I could cross out the 2, 2 and
4 to leave the prime 3.

Now consider whether it is possible to think of a prime number
in which none of its digits can be crossed out to make a new
prime. Note that digits cannot be rearranged, so in 971 you
are not allowed to form the prime 19. It turns out that it is
possible, but only twenty-six primes exist with this property.
They are called the minimal primes, and were found by
Jeffrey Shallit in 2000. The number 66600049 is the largest
of the minimal primes. Composite numbers have this property,
too. Think of any composite (non-prime) number. You can
always cross out digits to find another composite number,
except for a list of thirty-two minimal composite numbers, of
which 731 is the largest.

66000049

946669

666649

60000049

The five largest minimal primes.

70000000

The twin prime conjecture states that there are infinitely many pairs of primes separated by two; for example, 5 and 7, 11 and 13 or 17 and 19.

Polignac's conjecture states that there is nothing special about the number two, and that there should be infinitely many pairs of primes separated by any even number. For example, there should be infinitely many prime pairs separated by six, like 5 and 11. (These are called sexy primes, see page 66.)

Polignac's conjecture has not been confirmed for any value of n, but a breakthrough came in 2013. Yitang Zhang proved that there were infinitely many pairs of primes separated by some number n less than seventy million. The mathematical community, as part of the crowdsourced Polymath Project, have managed to reduce this upper bound from seventy million to 246. Assuming another conjecture about patterns in primes, the Elliott–Halberstam conjecture, the bound can be brought down to six.

Yitang Zhang's breakthrough set mathematicians on their way to proving that there could be infinitely many pairs of primes separated by an even number, and not just two.

73939133

This number is the largest right-truncatable prime number. This means that if you successively remove digits from the right, the remaining numbers are all prime. So 73 is prime, 739 is prime, 7393 is prime, and so on.

There are exactly eighty-three right-truncatable primes. Since all primes above five end in either a 1, 3, 7 or 9, these are the only digits allowed in such numbers.

It is also possible to find left-truncatable primes, in which removing digits from the left always gives a prime. These are more common than right-truncatable primes, and there are 4260 such examples, with the largest having twenty-four digits. Some smaller examples are 673, 1223 and 24967. If zeros are permitted there are infinitely many examples.

Fifteen primes exist that are both left- and right-truncatable, with the largest being 739397.

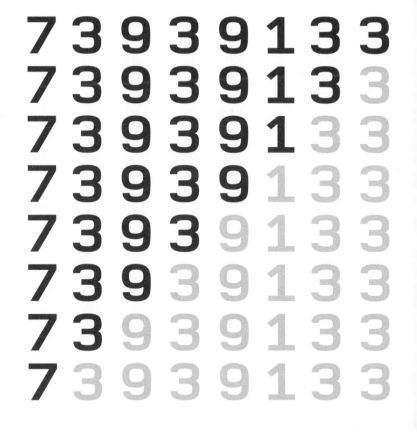

123121321

A permutation is an ordering of a set of symbols. For example, there are six permutations of the digits 1, 2, 3: 123, 132, 213, 231, 312 and 321. A superpermutation of a set of symbols is a string that contains every possible permutation. For example, you can make a superpermutation of 1, 2 and 3 by writing all six permutations one after the other: 123132213231312321. This has a length of eighteen digits, but it is possible to make shorter ones and the best solution here is 123121321, with a length of nine. This short solution makes use of overlaps between permutations – for example, 123 and 231 are both contained within 1231. So, for n symbols, how long is the shortest superpermutation? When n is 5 or less, there is a formula for the answer, which is $1! + 2! + \ldots + n!$, where $n!$ means the product of all the numbers up to n. For example, when $n = 3$ the answer is $1! + 2! + 3! = 1 + 2 + 6 = 9$. But for n bigger than 5, the formula fails. In 2011, Bogdan Coanda posted a new best result for $n = 7$ (length 5907) on YouTube, and the current best possible result stands at 5906.

```
3761254371625437126543712654371256431725463172543617254316725431762543172654317256431725463127
5436127543162754316275431276543127564312574361257436125743162574316275431276543125674312576431
2574631254763124576312475631246753124675731246543172446531724653172465531724653172453716245
3712645371246531724563172453617245316724531762453172645317246531724563172465317245631724531627
4531264745312764531247645312746531246735124673512463751243675124376512437651243751624375126437
2463571246531724356712435671246435716243571264357124365172436517243561724356172435716724351762435
1726435172463517243651724356172435162743516274351762743512674351726435172465317246315246731725463715243
6715243716152437165243715624371524631725436172543617245617254316755172436172465317255176451724354231764435
41356627413652741362527413627541362745136274513624751362451736245513762453137624513576241375
62413765241376254137624513672415367241356724135672413675241376524136725416372451632741563274126
53274163527416325741632754163274516324751632425716324513762456173245671732456173712451315256173
56732145763214576521457362145732614572361472453614725361472312536514723165147231654723651645726
31472563147263531472631547253147521361475263174526131745261374526173452613745531472715263457143
26143527614352671435261714532361714526173451261745261317452613745261714526137452613745617245
57261453726145732164573216475231647525316472531647235647235164723516472516472516472164752136472513
64721536472135564721365472163472163472156341721545723164721365372561472516427567136472513
16435271645321765431716425317642653174256314725534175231647524371462534176231475231745234746235
21674526147235421364725614355721642351647526135142631724673164215423164721653721432135713246614
5623671453273516327645231647235164723514672531647253146725314672513467251437625314275316271427536
14275316427531642573164257364142573615425731642567164257364142573615425731642567164253746253174651467
12351462735514267351427635142756142735614273514267325146723514267351427635142756164253746253174653174651467
4237516423751642357164235714236714235671423567124235671423567142356746235714726351747263514724236 51742356
17423516724235167423516742356174235614726354162534716235146726351456271462517654625316426351427 316536427
3154623715142637154262371425326617514652371465723154642315746231574623157462315746231574623165 17542317564
23175642231754262315746231574261574231657423564247235164723161574623157426315742631574261574 263157426315742631574263174562134
7652134675271346572123465726134567134526713452361734652713462531734652713463142135647172345 613472317564231756542
1766113245617634257136425713462513746251374625613745271346253174652137642513746213746251374625173642513 425
76132457163425713642573146251734625134726513472651734625134762534176251374625135712465314725 136427513742651374
25613742561534267135142371647521536271465723146425371642653174256314725534175231647524371462534176231475 2317452347462 35  (this is inconsistent)
```
```
Detail from the superpermutation for *n* = 7 posted by Bogdan Coanda in the
comments of a YouTube video. It contains every possible permutation of
the digits 1 to 7 and, at the time, was the shortest known such solution.
```

3816547290

Find a number that uses all ten digits 0 to 9 exactly once, so that the first *n* digits are divisible by *n*. So the first two digits should be divisible by two, the first three digits divisible by three, and so on. The only number that will solve this puzzle is 3816547290.

There are some useful tricks to be found in solving this puzzle. For example, numbers divisible by ten end in 0, so the last digit is 0. Numbers divisible by five end in 0 or 5, so the fifth digit must be 5. The even digits must be even, meaning that the odd digits must be odd. The first three digits must sum to a multiple of three if they are to be divisible by three.

The solution to this puzzle is an example of a pandigital number in which every digit 0 to 9 is used at least once. Pandigital numbers using each digit only once cannot be prime, since the digit sum will always be 45, making the number divisible by both three and nine.

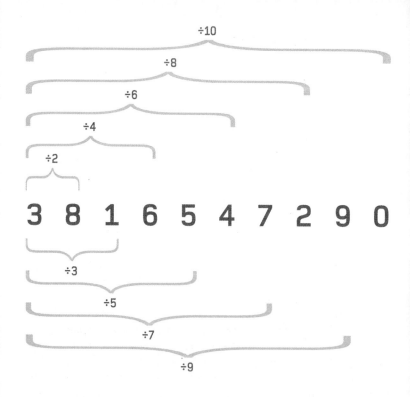

4294967297

This ten-digit number is F_5, the fifth Fermat number. A Fermat number F_n is a positive integer of the form $2^{2^n} + 1$ for some non-negative integer n. For example, $F_0 = 2^{2^0} + 1 = 2^1 + 1 = 3$, while $F_1 = 2^{2^1} + 1 = 2^2 + 1 = 5$. This means that F_5 is $2^{2^5} + 1$.

In 1650, Pierre de Fermat conjectured that all Fermat numbers were prime, based on his calculations of F_0 to F_4. Unfortunately for him, it turns out that F_5 is not prime, as it is the product of 641 and 6700417. This was shown by Euler in 1732. There are now two competing conjectures about Fermat numbers. One of them posits that there are infinitely many prime Fermat numbers, while the other suggests that all Fermat numbers above F_5 are composite. So far, 298 Fermat numbers have been shown to be composite, though only F_0 to F_{11} have been completely factorized. No Fermat numbers above F_4 have been found to be prime. Interested readers can join the distributed computing project *Fermat Search*, which uses spare processing power on people's computers to search for factors of Fermat numbers.

Pierre de Fermat

61917364224

This number is $144 \times 144 \times 144 \times 144 \times 144$ – that is, 144 raised to the fifth power. It is interesting because it can be written as the sum of four other fifth powers: $144^5 = 27^5 + 84^5 + 110^5 + 133^5$. As such, this number contradicts the 'sum of powers' conjecture by Euler, which states that we need at least three cubes to sum to another cube, at least four fourth powers to sum to a fourth power, at least five fifth powers to sum to a fifth power, and so on. The conjecture was an attempt to generalize Fermat's last theorem, which said that two nth powers cannot sum to another nth power for $n > 2$ (see page 82).

This counterexample to Euler's conjecture was found by mathematicians Lander and Parkin in 1966, after the conjecture had been open for nearly two hundred years. An infinite family of counterexamples was later found for the case of fourth powers, but only three counterexamples are currently known for the case of fifth powers.

$$61917364224 = 144^5$$

$$144^5 =$$
$$27^5 + 84^5 + 110^5 + 133^5$$

26534728821064

On a chessboard, a knight moves in an L-shape, counting two squares horizontally and one vertically or vice-versa. A knight's 'tour' is a sequence of moves in which a knight visits every square on the board exactly once. If the knight returns to its starting position, the tour is described as 'closed'. The picture opposite indicates one possible closed knight's tour on a standard 8 x 8 chessboard. But how many different tours are there?

If you count the number of 'directed' closed knight's tours (meaning the same path traced in the reverse direction is counted separately) then there are 26,534,728,821,064 different tours. Two paths that are a reflection or rotation of each other are counted separately as well. This problem is an example of a Hamiltonian path problem, which seeks a path through a graph, visiting every node exactly once. Generally, Hamiltonian path problems are NP-complete, meaning solutions can be checked quickly but there are no known fast methods for finding those solutions. However, the knight's tour problem can be solved in linear time (quickly).

4×10^{18}

Goldbach's conjecture is a famous unsolved problem concerning prime numbers. It states that every even number bigger than two can be written as the sum of two prime numbers. For example, 12 = 5 + 7; 18 = 5 + 13; 50 = 3 + 47. While nobody has yet proved that Goldbach's conjecture is true, it has been checked and found to hold for all even numbers up to four billion billion (4×10^{18}).

There is a variant of the conjecture, called Goldbach's weak conjecture. This states that all odd numbers bigger than five are the sum of three prime numbers. In 2013 Harald Helfgott produced a proof of Goldbach's weak conjecture. This is widely accepted but, as of printing, has not been published in a peer-reviewed journal. Helfgott's proof, if confirmed, would imply that every even number can be written as the sum of, at most, four primes. The next best confirmed result is by Olivier Ramaré, who showed, in 1995, that every even number is the sum of, at most, six primes.

$$4 \times 10^{18} =$$

$$400000000000000000000 =$$

four

billion

billion

9.2234×10^{18}

Whoever invented chess is said to have approached their ruler to reward them for their creation. They asked for wheat according to the following rule: one grain of wheat placed on the first square of the board; two grains of wheat on the second square; four grains on the third square; eight grains on the fourth square and so on, doubling with each square. The ruler laughed, wondering why the inventor had asked for such a meagre reward. But how much grain would there be on the chessboard at the end of the procedure?

An 8 × 8 chessboard has sixty-four squares. This means that the last square will contain 2^{63} grains of wheat, which is 9.2234×10^{18} (nine billion billion) grains. This is some 460 billion metric tonnes of wheat, which is around 600 times more than current world wheat production. The total number of grains on the chessboard is $2^{64} - 1$, which is calculated by summing all the powers of 2: $2^0 + 2^1 + 2^2 + \ldots + 2^{63}$. This is known as a geometric series.

1.8447 x 10^{19}

The Tower of Hanoi is a puzzle involving a set of discs of different sizes and three rods. With the discs arranged in size order on one pole, smallest at the top, the puzzle is to move all of the disks to another pole. But there are rules.

You can only move one disc at a time; a smaller disc can only be placed on top of a larger disc; and, because the discs are stacked on the rods, you can only move the top disc of a given stack each time.

The creator of the puzzle, Edouard Lucas, spoke of an ancient legend where monks had to move sixty-four discs in this way, after which the world would end. How long would this take? The number of moves required to solve the Tower of Hanoi with n discs is $2^n - 1$. This can be shown using a technique called proof by induction. By the same method, the monks would need $2^{64} - 1 = 1.8447 \times 10^{19}$ moves to solve the puzzle. Assuming one second for each move, this is nearly 600 billion years.

Tower of Hanoi

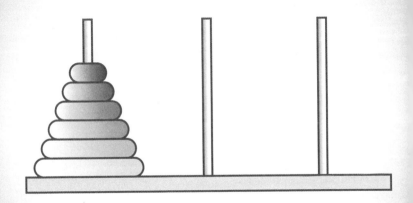

A tower with six discs can be solved in sixty-three moves.

43252003274489900000

This twenty-digit number, approximately forty-three billion billion, is the number of configurations of a standard $3 \times 3 \times 3$ Rubik's cube. It is calculated by considering the number of positions that each little cube can be in within the larger cube.

There are eight corner cubes, and each of these can sit in three possible positions. This gives us 8! = 40320 permutations multiplied by 3^8 choices for their orientations. There are twelve cubes that are edge pieces, and each of these cubes can sit in two possible positions. This gives us 12! = 479001600 permutations multiplied by a further 2^{12} choices of orientation. This gives you $8! \times 3^8 \times 12! \times 2^{12}$ = 519 billion billion combinations. However, some of these are impossible – for example, you cannot rotate a single corner piece by itself, so must divide the number by three. You cannot flip a single edge piece, so must divide the number by two. Finally, you cannot swap a single pair of edge pieces, so must divide by two again. This gives the answer of forty-three billion billion.

43252003274489900000 can also be written as 4.325 × 10^{19}.

6.6709 × 10¹⁹

‎

A sudoku puzzle is a special type of Latin square in which every row and column sum to the same total. In its classic form it is a 9 × 9 grid where the digits 1 to 9 must appear in each row and column, and also in the nine 3 × 3 subgrids. The puzzle setter must include enough digits to ensure the grid has one unique solution. But how many different sudoku grids are there? The answer for 9 × 9 grids is 6670903752021072936960, or 6.67 thousand billion billion (6.6709 x 10¹⁹). This was first stated by user 'QSCGZ' to the rec.puzzles newsgroup in 2003 and confirmed by mathematicians in 2005.

However, many of these are the same solution in disguise, being a rotation, reflection, permutation or relabelling of each other. When these symmetries are taken into account, there are about 5.5 billion essentially different solutions. In 2012, it was proven that a minimum of seventeen digits must be included for a 9 × 9 puzzle to have a unique solution. Any fewer than seventeen digits, and this is not the case.

5	3			7				
6			1	9	5			
	9	8					6	
8				6				3
4			8		3			1
7				2				6
	6					2	8	
			4	1	9			5
				8			7	9

8.0802 × 10⁵³

The Monster group is the largest of the sporadic simple groups, containing about 8×10^{53} elements. This is around one thousand times more than the number of atoms that make up our planet.

Groups are algebraic structures that capture our ideas about symmetry. Simple groups are to groups as primes are to numbers, in that they are the building blocks of all other groups. A simple group is one that, in some sense, cannot be divided into smaller groups. One of the largest proofs in all of mathematics was the classification of finite simple groups, which showed that every finite simple group was either a member of one of four families of groups, or it was one of twenty-six sporadic groups (see page 60). The sporadic groups do not seem to follow any pattern, though many of the twenty-six groups sit inside one another. The Monster group contains all but six of the others. There are other simple groups that are larger than the Monster, but the complicated structure of the Monster makes it the hardest to understand.

The Monster group

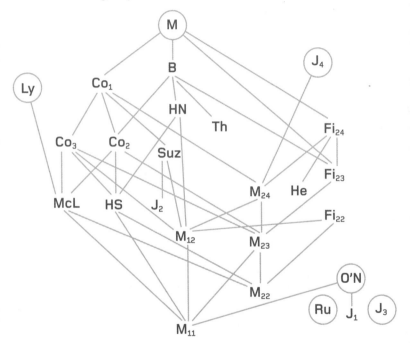

A diagram showing how the twenty-six sporadic simple groups relate to one another. The largest circles indicate the six groups that are not contained within the Monster group.

52! (52 factorial)

This incredibly large number – an 8 followed by sixty-seven digits, comparable to the number of atoms in the Milky Way – comes from a very simple everyday item: it is the number of ways of shuffling a standard deck of fifty-two cards.

How to calculate this number? Imagine choosing each card one by one and see how many possibilities there are for each. There are fifty-two choices for the first card. Once this is picked, there are fifty-one choices for the second card, then fifty for the third, and so on, decreasing by one each time. The total number of different shuffles is then the product of all these numbers: $52 \times 51 \times 50 \times 49 \ldots 2 \times 1$. This can be written in mathematical notation as 52!, called 52 factorial. The sheer size of this number means that any truly shuffled deck of cards has almost certainly never been seen before in the entire history of humanity. Even if a billion decks of cards had been shuffled every second since the Big Bang, this fact would still be true.

10^{100} (a googol)

A googol is a 1 with one hundred zeros after it. Other names for the googol are ten duotrigintillion or ten thousand sexdecillion. The name 'googol' was first written down by American mathematician Edward Kasner, who asked his nine-year-old nephew Milton Sirotta to pick a name for this giant number. This number is already bigger than the count of all the elementary particles in the known universe. But it is possible to go one step further with a number called the googolplex, which is a 1 followed by a googol number of zeros. If you consider the number of permutations of all the particles of the universe (including things like photons and force-carrying particles), then this is getting close to a googolplex.

Some readers may notice a similarity between the word 'googol' and the name of the search engine 'Google'. This is no coincidence: Google is apparently an accidental misspelling of this enormous number. The domain name 1e100.net (the scientific notation for a googol) is used by Google to identify its servers.

$10^{100} =$ 10,000,000,000,
000,000,000,000,000,
000,000,000,000,000,
000,000,000,000,000,
000,000,000,000,000,
000,000,000,000,000,
000,000,000,000,000.

101! + 1

T he gaps between consecutive prime numbers are hard to predict. Sometimes there is the smallest possible gap of two – for example, between 11 and 13 – and sometimes you have to wait longer for the next prime, such as between 113 and 127. But how big does the gap between consecutive primes get? And is there a 'biggest' gap?

The answer is 'no': for any even number you think of, there is a pair of consecutive primes with at least that gap. For example, after the number given here, which is roughly a 9 with 159 digits after it, there is a gap of at least 100 before the next prime. How did we find this number? It is equal to 101! + 1, meaning the product of all the numbers up to 101, plus one. Since 101! is equal to $2 \times 3 \times \ldots \times 101$, it is a multiple of all the numbers from 2 up to 101. This means that 101! + 2 is a multiple of two, 101! + 3 is a multiple of three, and so on up to 101! + 101 being a multiple of 101. This gives a run of one hundred consecutive numbers that cannot be prime.

101!+1

RSA-2048

Factorizing a number means breaking it into a product of smaller numbers. For example, you could factorize twelve as 3 × 4 or as 2 × 6 or even 2 × 2 × 3. The most difficult numbers to factorize are semiprimes – products of two prime numbers. For example, can you factorize 4843?

The difficulty of this problem makes it perfect for encryption: just as a lock is easy to close but hard to open, it is easy to multiply two primes together, but hard to factorize and get the numbers back again. This principle is used in RSA (Rivest–Shamir–Adleman), one of the most famous and widely used encryption systems.

In 1991, RSA Laboratories published a list of fifty-four large semiprimes and offered prizes of up to US $200,000 to people who could factorize them. The challenge ended in 2007 with only twelve of the numbers solved. The largest unsolved RSA number is RSA-2048, with 617 decimal digits (2048 bits).

Largest known prime

A prime number is a positive whole number with exactly two divisors: itself and one. It has been known since ancient times that there are an infinite number of primes; there will never be a largest one, but finding large primes is important both in number theory (finding patterns in primes) and in cryptography.

As of 2019, the largest prime number was $2^{82589933} - 1$ and has over twenty-four million digits. It is an example of a Mersenne prime, being one less than a power of two (see page 70), and was discovered by Jonathan Pace using the crowdsourced software GIMPS (Great Internet Mersenne Prime Search).

GIMPS uses spare processing power on computers to run an algorithm specially designed to detect Mersenne primes. The structure of these primes makes them easier to find than other types of primes. Nine of the ten largest known primes are Mersenne primes.

$$2^{82589933} - 1$$

The Electronic Frontier Foundation has a number of prizes for those who find large primes. There is $150,000 for the person who finds a prime with at least 100 million digits, and $250,000 for a prime with at least one billion digits.

Skewes's number

The prime number theorem states that the number of primes up to a value x is approximately $x/\ln(x)$. (Here $\ln(x)$ is the natural logarithm of x; see page 296). It predicts that the number of primes up to one hundred is roughly $100/\ln(100)$ = 21.7 (the true value is 25.) A hundred years after this result was discovered, mathematicians realized that the formula was acting as a crude estimate for the area below the $1/\ln(x)$ graph. By using calculus (specifically, integration, which divides an area into infinitesimally small rectangles) they were able to arrive at a better estimate of the prime number count. The resulting function (called li(x) for 'logarithmic interval') overestimates the true prime number count most of the time, and mathematicians predicted that it would overestimate all the time. However, in 1914, J. E. Littlewood proved that li(x) would flip between over- and underestimates infinitely often. In 1955, Skewes provided an upper bound for the first flip, shown opposite, now called the Skewes's number, though mathematicians have since found better bounds.

Skewes's number, as of 1955. Our current best estimate for Skewes's number is around 1.3×10^{316}, though it's possible a smaller one will still be found. A proof of the Riemann hypothesis (see page 378) would also help improve Skewes's number.

$$10^{10^{10^{964}}}$$

Graham's number

Graham's number became famous in the 1970s for being the largest number ever used in a mathematical proof. While other larger numbers have since been used, Graham's number still holds a place in the public imagination. This number is too large to be written down, even using every possible quantum unit of space in the entire universe. Even writing down how many digits it has cannot be expressed in the known universe.

Graham's number is an upper bound for a problem in Ramsey theory – an area of mathematics that studies order within large structures. The problem asks how many people need to be in a room so that you are guaranteed to find four people who all know each other or who are all strangers. (There is an additional constraint, that the four people must lie on the same plane if you imagine everyone as vertices of a hypercube.) The true answer may be as low as thirteen, or as high as Graham's number. Despite it being impossible to write out Graham's number, it is known that the last digit is 7.

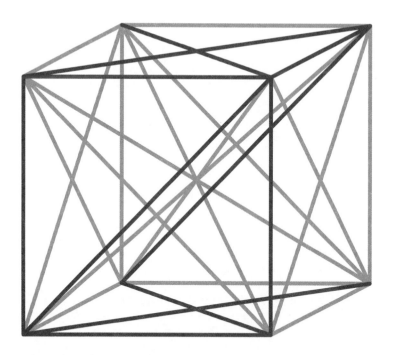

Take a cube and draw a line between every pair of corners. Colour each line one of two colours. Is it always possible to find four points so that all the lines joining them are the same colour? In the example shown here it is possible, but it is not always possible for three-dimensional cubes. Graham's number provides the dimension of cube for which the problem is guaranteed to work.

0.01 (1%)

According to the Global Rich List website at the beginning of 2019, an income of £25,300 ($32,000) was enough to put you in the top one per cent of the richest people in the world by income. But what does this mean? The words 'per cent' come from the Latin *per centum*, meaning 'by the hundred'. One per cent means 'one hundredth', or 1 in 100, or 0.01. To calculate 1% of a number, you divide it by one hundred. For example, if there are three billion people in the world earning money, being in the top one per cent of earners means 3 billion/100 = 30 million people who could earn more than you do (and 2.97 billion who earn less).

As their Latin origins suggest, percentages were first used in ancient Rome to specify the taxes people paid to the emperor. They are still used for that purpose today, as well as for interest rates, inflation rates (commodities are getting more expensive by about 2% per year) and general statistics (99% of people love maths!).

It is a common myth that the per cent sign comes from the two zeros in 100. It actually derives from the Italian *per cento*, which over time was abbreviated to pco, with the p being dropped and the co turning into %.

$^1/_{89}$

The number eighty-nine is both a Fibonacci number and a Fibonacci prime, but it has a much deeper relationship to the Fibonacci sequence. This is the sequence of numbers calculated by starting with 0 and 1, then adding the previous two numbers to get the next one, giving us the sequence 0, 1, 1, 2, 3, 5, 8, 13, 21, 34, 55, 89, . . . (see page 50).

Writing each of the Fibonacci numbers after the decimal point, and shifting the decimal point one place after each number, gives the sequence 0.0, 0.01, 0.001, 0.0002, 0.00003, 0.000005, 0.0000008, 0.00000013, etc. Adding up these Fibonacci fractions gives us the magic number of $^1/_{89}$.

This result can be shown using some algebra and the properties of the Fibonacci sequence, and we see eighty-nine in the answer because $89 = 100 - 10 - 1 = 10^2 - 10 - 1$. If, instead, we wrote out Fibonacci decimals by moving the decimal point two places after each number, the sum would be 1/9899, since $9899 = 100^2 - 100 - 1$.

.0112359550**5**

¹/₈₉ expressed as a decimal.

0.0167

Johannes Kepler revolutionized science when he suggested that the planets moved in ellipses around the sun, rather than circles as had always been believed. Not only did it allow astronomers to predict the motions of the heavens more accurately, but it laid the groundwork for Newton's theory of gravitation eighty years later.

So, what is an ellipse? If you draw two dots a little distance apart, then an ellipse is the collection of all points such that the sum of the distances to the two dots is a fixed number. The two dots are called the foci (singular: focus) of the ellipse. In our solar system, each planet moves on an elliptical path with the sun as one of the foci. The exact shape of each planet's orbit is a little different – some are more circular and some more elongated. This variation in shape is captured by a number between zero and one called the eccentricity of the ellipse. The Earth's orbit has an eccentricity of 0.0167, meaning it is very close to being circular.

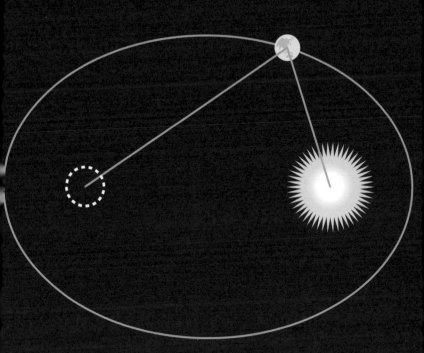

The Earth travels around an ellipse with the sun at one focus. The sum of the distances from the Earth to the two foci of the ellipse is always the same. However, the path of the Earth is not as elongated as in this diagram — it is very close to being a circle.

0.0458

If you take any newspaper and list all the numbers that appear in it, you would expect the leading digit to take the values 1 through to 9 equally often. Yet this is not the case, according to a phenomenon known as Benford's law.

Benford's law states that the distribution of leading digits in real-life numerical data is logarithmic. More precisely, the probability of the leading digit being d is $\log_{10}(1 + 1/d)$. For example, the chance that the leading digit is a 9 is $\log_{10}(1 + 1/9)$ $\approx 0.0458 = 4.6\%$. This is much lower than our intuition of 11%, and much lower still than the corresponding value for the digit 1, which is around 30.1%. Benford's law works best with data that span a number of magnitudes. So it would work well on data about sizes of lakes, but less so on shoe sizes or human heights. Fraudsters unaware of Benford's law usually falsify data with each leading digit being equally common, making it easy for police to detect. It has been used in this way to detect financial, economic and scientific fraud.

Benford's law

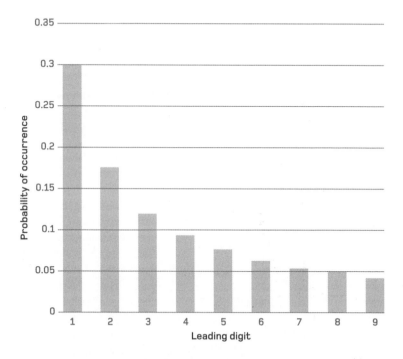

The way that leading digits are distributed among numerical data is not even, but instead follows the logarithmic shape seen here.

0.110001...
(Liouville's constant)

Every number is either rational, meaning it can be expressed as a whole-number fraction $^a/_b$, or irrational, meaning it cannot. Examples of irrational numbers include π, $\sqrt{2}$ and φ (the golden ratio; see page 328). But some irrational numbers are deemed 'more irrational' than others. Any irrational number can be approximated by rational numbers. For example, π is roughly equal to 3 (see page 350). A better approximation is $^{22}/_7$ (see page 352), and yet another is $^{355}/_{113}$. In general, the bigger the denominator, the more accurate the approximation. The 'irrationality measure' of a number assesses how well a number can be approximated by fractions, relative to the size of the denominators. The bigger the measure, the better the approximations. Under this measure, φ is more irrational than π. There is also a whole class of irrational numbers, called Liouville numbers, that are barely irrational at all: their irrationality measure is infinite. The Liouville constant was the first such number to be discovered. It was also the first number proven to be transcendental, meaning it is not the solution of any equation using only powers of x and whole numbers, like $x^2 + 2x - 1$.

$$L = \sum_{n=1}^{\infty} 10^{-n!} =$$

0.1100010000000000000000000100

Liouville's constant is the decimal fraction created by putting a 1 in each decimal place corresponding to a factorial $n!$ and zeros elsewhere. (The factorial $n!$ is calculated by multiplying all the numbers up to n—for example, $3! = 3 \times 2 \times 1$.)

0.12345678910... (The Champernowne constant)

The Champernowne constant is the decimal number created by writing all the positive whole numbers in order after the decimal place. There is no repeating pattern in its digits, so this number is irrational, meaning it cannot be written as a whole-number fraction. Furthermore, it is transcendental. This means that it is not the solution of any equation involving integers and powers of a variable x.

A more interesting property of the Champernowne constant is that it is 'normal'. This means that every digit, and every sequence of digits, occurs equally often. So each of the single digits 0 to 9 appears 10% of the time; each of the two-digit numbers 00 to 99 appear 1% of the time, and so on. Mathematicians have shown that, in some sense, 'most' infinite decimals are normal, but finding normal numbers is difficult. The constants π and e are suspected to be normal, but it is not known whether every digit occurs infinitely often or whether every possible sequence of digits can be found.

0.12345678910111213
14151617181920 2122
23242526272829 303
13233343536373839
40414243444546474
84950515253545556
57585960616263646
56667686970717273 7
47576777879808182 8
38485868788899091
929394959697989 9 ...

0.2079

The number *i* is an imaginary number (see page 394). Equal to the square root of −1, it does not sit anywhere on our familiar number line. Yet, quite amazingly, if we raise *i* to the power of *i*, we get a real number, roughly equal to 0.208! But what does this calculation mean?

With real numbers, the expression '*a* to the power of *b*', or a^b, means multiplying *b* copies of *a* together. For example, 2^3 is equal to $2 \times 2 \times 2$. So what does it mean to multiply *i* copies of *i* together? Since the question does not make sense when phrased this way, mathematicians seek to find other routes to an answer.

Imaginary numbers create a relationship between the constant *e* (see page 346) and the trigonometric functions sine and cosine. The formula, called Euler's formula, is as follows: $e^{ix} = \cos(x) + i \sin(x)$. Here *x* is an angle measured in radians (see page 366). When we set *x* equal to $\pi/2$ (or 90°), we get $e^{i\pi/2} = \cos(\pi/2) + i \sin(\pi/2) = i$. Raising *i* to the power of *i* now gives us $(e^{i\pi/2})^i = e^{-\pi/2} = 0.2079$.

$^2/_9$

The Cantor set is a remarkable mathematical object, full of contradictions that challenge our intuition about the number line and its geometry. Draw a straight line of length 1. Now divide your line into three equal segments and erase the middle one. Repeat the procedure with each of your two remaining lines. Continue doing this to each line segment that is left, to create a series of more and more, but smaller and smaller, line segments. Now imagine that the drawing goes through infinitely many iterations of dividing and deleting. What you will have created is a fractal – a shape that looks the same no matter how far you zoom into it. So, what proportion of the original line have you deleted? The answer to this is an infinite sum ($^1/_3$ + $^2/_9$ + $^4/_{27}$ +...), which turns out to have the answer of 1. So it seems you have deleted all of it. Yet, in a wonderful contradiction, you can also see that (uncountably) infinitely many points are not deleted. One such example is the point $^2/_9$ of the distance along the line. It is mind-boggling that you can end with as many points as you started with, yet they take up no space at all.

The Cantor set

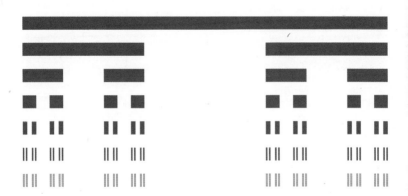

Divide a line into three equal parts and delete the middle part. Repeat on each remaining line segment. The points that remain after this infinite procedure form the Cantor set (which, incidentally, was not discovered by Cantor, but by Henry George Steven Smith).

0.3010

The logarithm of two, written log(2) or $\log_{10}(2)$, is the power to which ten must be raised to get an answer of two. That is, it is the solution of the equation $10^x = 2$. It is an irrational number with an approximate value of 0.301.

John Napier popularized logarithms in the seventeenth century, valuing them because they could be used to simplify arithmetic. Logarithms turn multiplication problems into addition problems (and division into subtraction). This is because of the formula log(ab) = log(a) + log(b), which equates the logarithm of a product with the sum of individual logarithms. Log tables and slide rules were primary tools for arithmetic until pocket calculators became widely available in the 1970s. These devices allowed people to switch quickly between numbers and their logarithms, making it easy to do large multiplications. Today we no longer use logarithms for calculation, but they have many applications in science and engineering. Logarithmic scales are used to measure sound levels, earthquakes, acidity and star brightness.

John Napier

Engraved by G. Freeman

$^1/_3$

Sometimes, when a fraction is written as a decimal, it requires infinitely many decimal places. The fraction $^1/_3$ is the simplest example of this, as its decimal representation is 0.3333… with infinitely many threes. This issue is not specific to our use of a decimal (base-10) number system, as it occurs in every positional number system. In the Babylonian base-60, for example, $^1/_7$ has an infinite representation, although $^1/_3$ and $^1/_6$ are finite.

The question as to which numbers can be written in a finite way in a positional system relates to the prime factors of the base. In the case of decimals, these are two and five. Therefore the only fractions with finite decimals are those whose denominators are a product of only twos and fives. For example, $^1/_8$ = 0.125 is finite because 8 = 2 × 2 × 2, and $^3/_{20}$ = 0.15 is finite because 20 = 2 × 2 × 5. The number 60 has prime factors 2, 3 and 5, so any fraction whose denominator is a product of twos, threes and fives will have a finite representation in base-60. This makes many fractions much easier to work with in base-60 than in base-10.

0.333333...

The fraction ⅓ written as a decimal, with infinite decimal places.

0.3405 (34%)

Should you go out tonight or stay in and read a thrilling maths book? Perhaps toss a coin to make your choice. And what if every decision in life was decided in this way? Could we make any predictions about how it would turn out? The answer is, sometimes, yes. Randomness can be predictable. A simple example of this is called a random walk, where you toss a coin to decide which direction to move in. In the one-dimensional case you move a step forwards if you toss heads, a step backwards if you toss tails. Mathematicians can prove that you will always end up back where you started if you keep moving this way. For the two-dimensional case, picture a drunken walk through a city where, at each intersection, you randomly pick one direction to go in. Maths once again predicts that you will eventually end up back where you started. Yet, in three dimensions, a random walk has only a 0.3405 (34%) chance of getting you back to the start. Simulations of these random walks are important in studying the motions of air particles, of neurons firing in our brains and even estimating the size of the World Wide Web.

A random walk through
a city after 2500 steps,
choosing a random
direction at each
intersection.

0.3679 ($^1/_e$)

This number is $^1/_e$, where $e = 2.718$ is the famous constant of calculus. It is used to solve a problem called the marriage problem or secretary problem.

Imagine a situation in which Alice has one hundred suitors asking for her hand in marriage. She will speak to each one in turn, and after each meeting must decide whether to accept or reject the proposal. We assume that the candidates can be ranked from best to worst and that the order in which they are interviewed is random. If Alice rejects a suitor, she cannot change her mind later. Mathematicians have worked out that the optimal strategy for Alice is to interview and reject the first 36.79% of the candidates and then to marry the next person who is better than all the suitors she has seen so far. This strategy gives a 36.79% probability of choosing the best person, which is better than any other method. The magic number $^1/_e$ appears because calculus is needed to estimate where the best suitor is in the queue.

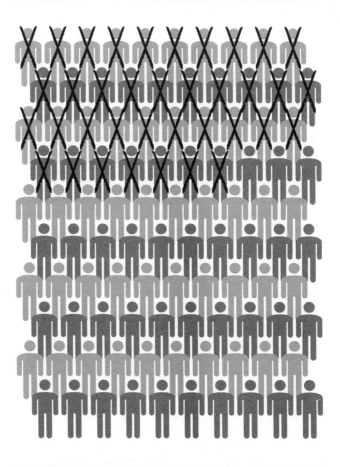

0.4124

How should two people choose items from a pile to get the fairest distribution, when taking turns gives an unfair advantage to the first person? A seemingly unrelated puzzle is to divide the numbers zero to fifteen into two groups so that the sums of each group are the same, the sums of the squares in each group are the same and the sums of the cubes in each group are the same. Both puzzles are solved using a wonderful sequence called the Thue-Morse sequence. Write each whole number in binary (see page 12). The digits 0 to 5 are written in binary as follows: 0, 1, 10, 11, 100, 101. For each binary number in turn, call it odious and assign it a '1' if it contains an odd number of ones. Call it evil and assign it a '0' if it contains an even number of ones. The Thue-Morse sequence is the resulting sequence of 0s and 1s: 01101001.... To solve the fair division problem make one person 0 and the other 1. Taking turns as the sequence suggests is the fairest way to share the goods. For the maths puzzle, put the evil numbers in one group and the odious numbers in the other group, and the sums of the numbers, their squares and cubes will match.

What is the best way to arrange an eight-person rowing team? If the rowers alternate between left and right sides, the boat will end up with a slight sideways imbalance. The Thue-Morse sequence, which starts 01101001, gives the answer: rowers 1, 4, 6 and 7 go on one side, while 2, 3, 5 and 8 go on the other side. Turning the number 0.1101001… from binary to decimal gives 0.4124.

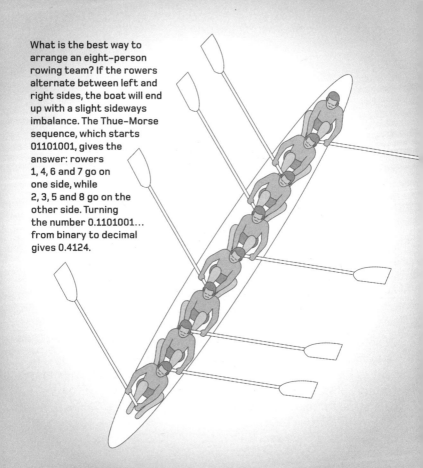

$^1/_2$

When you halve something, you divide it into two equal parts. Each part is said to be one half of the whole, written in fractions as $^1/_2$, in percentages as 50% or in decimals as 0.5. In probability, a chance of 50% means the event is just as likely to happen as not.

The Greek philosopher Zeno of Elea, writing in 450 BCE, came up with a baffling paradox using the number $^1/_2$, showing that motion was impossible. 'To cross a room,' he said, 'I must first cross halfway. But to cross halfway, I must cross a quarter of the way, and before that one-eighth of the way. In fact, I must do an infinite number of tasks before I can begin to move!'

Zeno's dichotomy paradox, as it is known, is usually either resolved using modern techniques of infinite sums (showing that $^1/_2 + ^1/_4 + ^1/_8 + ^1/_{16} + \ldots$ is genuinely equal to 1) or by arguing through physics that there is a smallest unit of length that cannot be further divided into two.

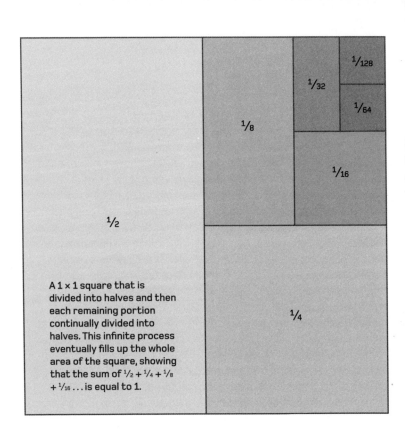

1/128

1/32

1/64

1/8

1/16

1/2

1/4

A 1 × 1 square that is divided into halves and then each remaining portion continually divided into halves. This infinite process eventually fills up the whole area of the square, showing that the sum of 1/2 + 1/4 + 1/8 + 1/16 ... is equal to 1.

0.506

Edward Lorenz was an American mathematician and meteorologist. In 1961, he was using a computer to simulate the weather, inputting variables such as temperature, pressure and wind speed, and letting the machine solve the complicated equations. One day, wanting to save time when repeating a calculation, he started the program from the middle, using the input 0.506, which the computer had printed the day before. To his surprise, Lorenz found that his simulation produced completely different results from the previous day. Eventually he realized that the problem was not the computer, but the printout. The number it had printed out was 0.506, yet the number it stored in its memory was 0.506127. Those extra three decimal places were enough to completely change the calculations. Lorenz had stumbled upon a mathematical phenomenon called chaos theory. Chaos here does not mean 'random' or 'disordered', but rather a system that is very sensitive to initial conditions. This means that a small change in what goes in leads to a large change in what comes out.

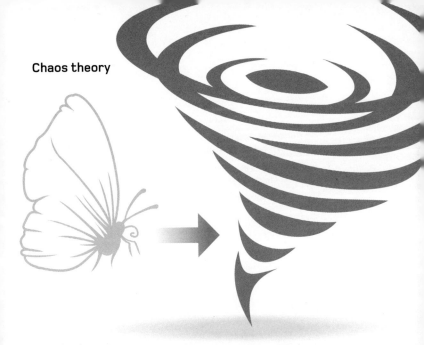

Chaos theory

The mathematical field of chaos theory is better known in popular culture as the 'butterfly effect'. The idea is that a butterfly flapping its wings in Brazil may cause a tornado in Texas – or, in mathematical language, a small change in initial conditions can produce a large change in output. The fact that weather is chaotic means that we will never be able to make predictions very far ahead, no matter how powerful our supercomputers are. This is not because we do not understand the equations, but because our measurements will never be completely precise.

0.56 (56%)

The prisoner's dilemma is a counterintuitive thought experiment in an area of mathematics called game theory. Two prisoners are arrested on a minor charge. The police want to convict them for a more serious crime, but need a confession to get enough evidence. The prisoners are separated and offered a deal: if they both stay silent they each get one year in jail for the minor crime; if they both confess to the major crime, they each get three years in jail; if one confesses and the other does not, the one who confesses is set free and the one who stays silent gets five years in jail.

What should the prisoners do? Each will reason as follows: 'If my partner stays silent, I get a better deal by confessing (0 vs 1 year). If my partner confesses, I also get a better deal by confessing (3 vs 5 years). I should therefore confess.' In this way both prisoners will confess and get three years in jail, yet if they had both stayed silent they would have got only one year. In a real-life experiment on prison inmates in 2013, researchers Khadjavi and Lange found that 56% of them cooperated with each other and stayed silent.

The prisoner's dilemma

The fact that 56% of prisoners stayed silent in a real-life experiment shows that rational mathematical analysis is not always a good predictor of human behaviour.

0.6602 (C_2)

The twin prime constant, denoted by C_2 or π_2, is approximately 0.6602 and is used in the Hardy-Littlewood conjecture about the distribution of twin primes. Twin primes are prime numbers separated by two – for example, 11 and 13 or 41 and 43. It is conjectured that there are infinitely many pairs of twin primes (see page 338). But how dense are the twin primes within the number line? How often do we find such pairs?

The number of all primes up to x is approximately $x/\ln(x)$. The Hardy-Littlewood conjecture (which remains unproven) states that the number of twin primes up to x is approximately $2C_2 x / (\ln(x))^2$.

The formula for C_2 is to multiply together all numbers of the form $p(p - 2)/(p - 1)^2$ where p is prime. This is an infinite product, but the terms get closer and closer to one: for example, $(101)(99)/100^2 = 0.9999$. This means the calculations converge to a precise number, which is 0.6602.

Hardy–Littlewood conjecture

$$\prod_{\substack{p>2 \\ p \text{ prime}}} \frac{p(p-2)}{(p-1)^2} = \frac{3 \cdot 1}{2^2} \times \frac{5 \cdot 3}{4^2} \times \frac{7 \cdot 5}{6^2} \times \frac{11 \cdot 9}{10^2} \times \cdots$$

The Monty Hall problem is a probability puzzle, named after the host of an American game show. In front of you are three doors. Behind one door is a car and behind the other two are goats. You select a door, after which Monty opens a different door to show you a goat. To win the car, are you better off sticking with the door you first chose, or switching to the other unopened door, or does it make no difference?

Most people, when seeing this problem for the first time, will say that it makes no difference whether you stick or switch. There are two remaining doors, and it seems equally likely that the car is behind either one. Yet the true answer is that you have double the chance of winning if you switch doors. When you make your first choice of door, you have a one-in-three chance of picking the car. Monty's goat reveal does not change this, but it does give new information about the remaining door. If you chose wrongly the first time (which happens two out of three times) then you are guaranteed to win the car if you switch.

The Monty Hall problem is sometimes called Marilyn and the Goats. This is because it became famous in 1990 in a reader's letter to the 'Ask Marilyn' column of *Parade* magazine. Marilyn vos Savant, famous for having the highest IQ in the world, replied to the reader explaining that the answer was $2/3$. Around 10,000 people wrote in to tell her she was wrong, including many people with PhDs in mathematics.

0.682 (68%)

My local supermarket sells lemons at the same price for each, yet they are not all the same size. If the supermarket asks farmers to ensure that, on average, each lemon weighs 40g, the farmers can still provide lemons with a range of weights. What is needed here, is a measure of how the weights of the lemons deviate from the average. The supermarkets can then ask the farmers to keep this deviation as small as possible. Instead of looking at the raw differences of each data point from the average, statisticians look at their squares. This has two advantages: negative and positive differences are both counted equally, and larger differences carry more weight. The variance is defined to be the average of the squares of the differences, and the standard deviation is the square root of the variance. With data that is normally distributed (like the bell curve opposite), 68% of the values will fall within one standard deviation of the average. Now, if the supermarket asks for the lemons to be, on average, 40g with a standard deviation of 2g, then 68% of the lemons will be within 38–42g.

The standard deviation

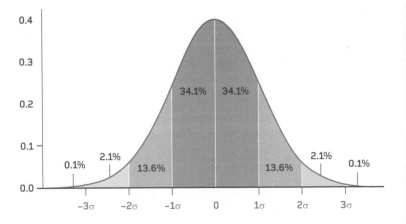

The standard deviation, σ, is a measure of the spread of a data set. Under a normal distribution, or bell curve, 68% of the data is within one standard deviation of the mean (average), 95% is within two standard deviations, and 99% is within three standard deviations.

0.6928

The Koch snowflake is a fractal, which means it is self-similar (repeats the same pattern) on every scale. It was one of the first fractals to be described, by Helge von Koch in 1904.

The first step in creating a Koch snowflake is to draw an equilateral triangle (all sides the same length). On each side, divide the line into three equal pieces and draw a new equilateral triangle using the middle segment as its base. Then delete the middle segment of the original line. Now repeat the procedure on each new side of the shape.

When this process is iterated infinitely many times, you obtain the Koch snowflake. At each step the perimeter is increased by a factor of $4/3$. This means that the Koch snowflake has infinite perimeter. Yet, amazingly, it has a finite area. If the triangle in the first step of construction has side lengths of 1, the area of the Koch snowflake is $\frac{2\sqrt{3}}{5}$, or 0.6928. This is $8/5$ the area of the original triangle.

How to draw a Koch snowflake.

0.6931

The natural logarithm of two, written ln(2), is the number to which the mathematical constant $e \approx 2.718$ must be raised to get the answer of two. That is, $e^{\ln(2)} = 2$. The answer is irrational and transcendental (see page 264), with an approximate value of 0.6931. Any process in which quantities are halved or doubled is likely to be described with a formula using ln(2). For example, the half-life of radioactive decay. This is the amount of time taken for a radioactive substance to decay to half its initial mass. It is used in cancer treatments, carbon dating and power stations. If the decay rate is λ then the half life is $\ln(2)/\lambda$.

Mathematically, ln(2) is interesting because it is the limit of the alternating harmonic series. That is, we alternately add and subtract the numbers $\frac{1}{1}$, $\frac{1}{2}$, $\frac{1}{3}$, $\frac{1}{4}$, etc. It is known as a conditionally convergent series, since the same sum with all the terms positive is infinite. The Riemann series theorem says that the terms of any conditionally convergent series can be rearranged to make the sum be any answer we want.

Alternating harmonic series

$$\sum_{n=1}^{\infty} \frac{(-1)^{n+1}}{n} = 1 - \frac{1}{2} + \frac{1}{3} - \frac{1}{4} + \frac{1}{5} - \frac{1}{6} + \cdots = \ln(2)$$

0.7048

A circle is a shape of constant width. This means that every diameter has the same length. Amazingly, the circle is not the only two-dimensional shape to have this property. The Reuleaux triangle is another such shape.

The term 'diameter' is slightly complex here. It means the distance between any pair of parallel lines that each touch the edge of the shape at least once, but do not go inside it. The top and bottom edges of the square opposite form such a pair of parallel lines for the Reuleaux triangle, as do the left and right sides. There are infinitely many shapes of constant width equal to 1, but the Reuleaux triangle is the one with the smallest area, equal to 0.7048. To construct this shape, start with an equilateral triangle (all sides equal). An edge of the Reuleaux triangle is then the arc of a circle centred on one vertex and joining the two other vertices. Several currencies have coins with shapes of constant width, an important property for use in vending machines. Reuleaux triangle drill bits are used to create square holes.

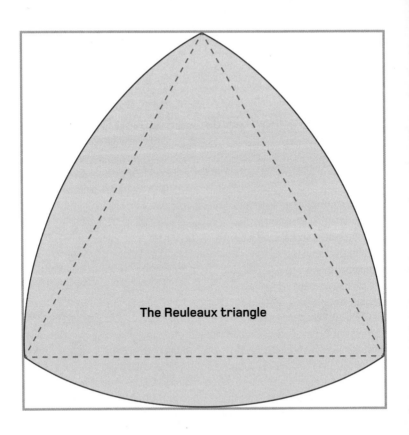

The Reuleaux triangle

0.7405 (74.05%)

In the sixteenth century, the explorer Sir Walter Raleigh wondered whether he was stacking his cannonballs in the most efficient way. He asked the astronomer onboard his ship, Thomas Harriot, who in turn asked his friend Johannes Kepler. Everyone agreed that Raleigh's best option was to arrange a bottom layer of cannonballs, and to place the balls of a second layer in the hollows between three lower balls. This is called hexagonal close packing and takes up 74.05% of three-dimensional space. Whether this really was the best packing solution became known as the Kepler conjecture and remained unsolved for nearly 400 years.

By the mid-nineteenth century, Carl Friedrich Gauss had proved that Raleigh's cannonball packing was the best one if the balls had to be placed in a regular symmetric way. But it was still possible that some haphazard way of arranging the spheres might be more efficient. In 1998, Thomas Hales presented a complete proof of the conjecture and it then took until 2017 before the proof was checked and fully accepted by the mathematical community.

Hexagonal close packing

Configuration of
the bottom layer

Thomas Hales produced his proof of the Kepler conjecture by writing
computer code to check over 5000 possible configurations.

¾ (3:4)

L issajous curves are created using a mechanical device called a harmonograph. In this device, a pendulum swings a pen back and forth in one direction, while another pendulum swings the paper back and forth along a perpendicular direction. The ratio of the frequencies of the two pendulums is the main factor that determines the shape of the curve. Whole-number ratios, such as 3:4, produce closed curves, where the end of the loop connects back to the beginning. Varying the phase of the pendulums can produce further figures: circles and ellipses result when the pendulums have the same frequency but are out of phase, for example.

As well as being beautiful to artists and geometers, Lissajous curves are important in engineering and astronomy. Sound engineers use them to check for the phase relationship between left and right audio channels, while many spacecraft travel around our solar system on Lissajous orbits in order to maintain a stable position with minimal propulsion.

Lissajous curves

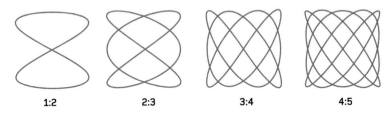

| 1:2 | 2:3 | 3:4 | 4:5 |

Four Lissajous curves created by changing the frequency
ratio of two pendulums in a harmonograph.

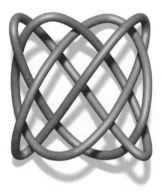

Lissajous knots are created by the combined motion of three
perpendicular pendulums. The frequency ratios for this knot are 3:4:7.

0.765

There are many different shapes that can tesselate (tile) three-dimensional space. This means they fit together exactly with no overlaps or gaps. Cubes are one example.

In 1887, Lord Kelvin questioned which three-dimensional tessellation was the most efficient – that is, which shape tiles three-dimensional space with the largest volume relative to its surface area? The corresponding question in two dimensions is the honeycomb conjecture (see page 358). Kelvin not only posed the question, but proposed a solution: the truncated octahedron. This is an octahedron with the corners sliced off, resulting in a symmetric shape composed of six squares and eight hexagons. The shape remained the best solution to Kelvin's problem until 1993, when two physicists found a better solution. Their Weaire-Phelan structure has an isoperimetric quotient of 0.765 (the ratio of its volume to that of a sphere with the same surface area). The shape beat Kelvin's by around one per cent. It is still an open question whether a more efficient shape might yet be discovered.

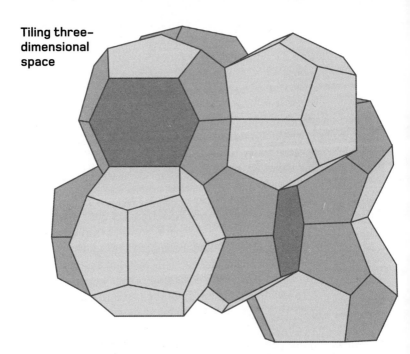

Tiling three-dimensional space

The Weaire–Phelan structure is made up of two different kinds of shape. One is an irregular dodecahedron, having all pentagonal faces, while the other has two hexagonal and twelve pentagonal faces. This structure inspired the design of the Beijing National Aquatics Centre for the 2008 Olympics, where its efficient shape meant it covered a large volume using the smallest amount of building materials.

$^4/_5$

—

We are comfortable working with fractions composed of any two whole numbers, such as $^4/_5$ or $^{13}/_2$, but ancient Egyptians preferred only to use fractions with a one in the numerator. For example, they would write the fraction $^4/_5$ as the sum $^1/_2 + ^1/_5 + ^1/_{10}$.

Writing in 1202, Italian mathematician Fibonacci described a greedy algorithm that would write any fraction as an Egyptian fraction – greedy, because it chooses the smallest denominator and thus the largest fraction at each step. When the fraction has either a two or three in the numerator (that is, any of the form $^2/_n$ or $^3/_n$), then Fibonacci's method gives an Egyptian fraction using at most three terms. It is still an open question, called the Erdős–Straus conjecture, whether any fraction with a four in the numerator can be written as an Egyptian fraction with three terms (as above for $^4/_5$). Computers have checked the result for all numbers $^4/_n$ where n goes up to 10^{17} (one hundred million billion) but a proof for all numbers remains elusive.

Egyptian fractions

This image shows the fraction $4/5$ written as an Egyptian fraction.
The eye-shaped symbol is drawn above a number to represent its reciprocal
– that is, one divided by that number. Since a vertical line means 1 and an
upside-down U shape means 10, this image represents $1/2 + 1/5 + 1/10$, which
is $4/5$. The unsolved question is whether every fraction of the form $4/n$ can be
written in this way using three numbers.

0.8927 (89.27%)

When it comes to tiling your bathroom, there are some shapes that will fit together perfectly, and others that will leave gaps. For example, triangles, squares and regular hexagons will tile your wall without gaps, while pentagons, heptagons and octagons will leave ugly wall showing behind.

It is still an open question as to which shape leaves the biggest gaps. Mathematicians restrict the question to those shapes that are convex, meaning shapes in which all internal angles are less than 180°. A first guess might be that circles are the worst at tiling, but this is not the case. Circles packed in the most efficient way take up 90.69% of space while the best packing of regular octagons is ever-so-slightly worse at 90.61%. But the current worst known tiling shape is the regular heptagon (seven sides), covering 89.27% of space. It is unknown whether there might be a more efficient way of tiling heptagons, or if there might be another shape whose best tiling is worse than this.

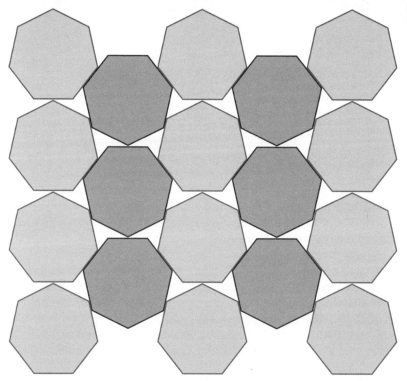

This is the current best guess at the most efficient way to tile a regular heptagon —a seven-sided shape whose sides are all the same length.

0.9999…

A fierce and very counterintuitive debate surrounds
the value of 0.9999… (with infinitely many nines).
Mathematicians say it is exactly equal to one, while most
people view it as a process rather than a number – something
that gets closer and closer to one without ever actually
reaching it. It is almost impossible for the human mind to
conceive of there being an infinite string of nines, rather than
a large but finite quantity. Ask people 'How close is 0.999…
to one?' and a common answer will be '0.000…1 – an infinite
number of zeros with a one at the end'. Despite the answer
not making mathematical sense, it seems more believable than
that 0.999… should be equal to one. Consider a mathematical
argument based on algebra. Let $x = 0.999…$, and multiply by 10
to get $10x = 9.999…$ Note that $10x$ has just as many nines as x
does. Now calculate the difference between the two numbers.
On the left-hand side you get $10x - x = 9x$, and on the right-
hand side you get $9.999… - 0.999… = 9$, since everything after
the decimal point cancels. Since $9x = 9$, you are left with $x = 1$.

$^{12}\sqrt{2}$ (1.0595)

Legend has it that Pythagoras, in passing a blacksmith's shop one day, noticed that the notes produced by certain hammers sounded good together. He then found that the most harmonious sounds came from hammers whose lengths were whole-number ratios of each other. This idea became the basis for tuning musical instruments using a system called 'just intonation'. For example, the notes C and G have a frequency ratio of 3:2 (called a perfect fifth), while consecutive C notes are in the ratio of 2:1 (an octave). The problem with just intonation is that the frequency interval between adjacent pairs of notes varies. For example, the step from B to C is different from the step from C to D. Since the eighteenth century, Western music has chosen a different tuning system, called 'twelve-tone equal temperament', where every interval is equal, while preserving the 1:2 ratio of octaves. This presents a mathematical puzzle: how to choose the frequencies of notes so that after every twelve notes, the frequency has doubled? The answer is that adjacent notes have a frequency ratio of $1{:}^{12}\sqrt{2}$ (the twelfth root of 2), which is about 1:1.0595.

Today's pianos are tuned using a system called 'equal temperament', where the frequencies of adjacent notes are always in the same ratio. This makes pianos sound reasonably good in every key, rather than being better in some keys than others. Music by older composers would have sounded different to their listeners, as their instruments were tuned differently.

1.202 (Apéry's constant)

The reciprocal of a number n is the fraction $\frac{1}{n}$. Mathematicians have long investigated what happens when the reciprocals of certain kinds of numbers are added together. For example, if you add up the reciprocals of the natural numbers ($\frac{1}{1} + \frac{1}{2} + \frac{1}{3} + \frac{1}{4} + \ldots$) the answer is infinite. The same is true for the reciprocals of the prime numbers. The sum of the reciprocals of the square numbers ($\frac{1}{1} + \frac{1}{4} + \frac{1}{9} + \frac{1}{16} + \ldots$), however, is finite (see page 330), as is the sum of the reciprocals of any higher powers. Adding up the reciprocals of the cubes gives a number called Apéry's constant, also known as $\zeta(3)$ or 'zeta of 3'.

Despite its relatively simple definition, Apéry's constant remains a mysterious number. It took until 1978 before anyone could prove it was irrational (meaning it can't be written as a whole-number ratio), and there is still no simple formula for it, in contrast to the sum of the reciprocal of the squares, which is $\pi^2/6$. If you pick three random whole numbers, the chance of there being no number that divides exactly into all of them is 1 in $\zeta(3)$, about 83%.

$$\zeta(3) = \frac{1}{1^3} + \frac{1}{2^3} + \frac{1}{3^3} + \frac{1}{4^3} + \frac{1}{5^3} + \dots$$

Apéry's constant is named after the French mathematician Roger Apéry, who proved that the number is irrational. It appears in quantum mechanics in calculations relating to electron spin, but nobody knows if there is a simple formula to calculate it.

$^3\sqrt{2}$ (1.2599)

In ancient times, there was a plague on the Greek island of Delos. The citizens asked advice from the oracle at Delphi, who instructed them to double the size of their altar to Apollo. The altar was cube shaped. The people doubled the length of each of its sides, yet the plague did not end. So, the Delian citizens asked the mathematician Plato for help. Plato said they should make a new cube with twice the volume of the original one. Doubling the lengths of each side, as the people had done, would create a volume eight times bigger ($2 \times 2 \times 2$), which was no good. The side length they needed was the cube root of two, written $\sqrt[3]{2}$, but what was its exact value? The problem of doubling the cube, also called the Delian problem, lingered for the next 2000 years. It is straightforward to find a numerical answer to the problem through trial and error – 1.2599 – but the Greeks wanted to find a purely geometric solution. This meant constructing a line of length $\sqrt[3]{2}$ using only an unmarked ruler and compass. In 1837, French mathematician Pierre Wantzel proved that the problem was impossible.

The Delian problem

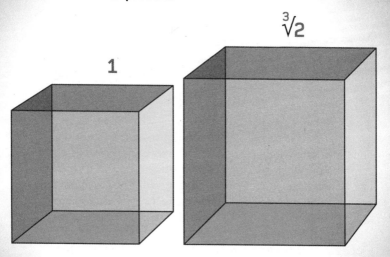

To double the volume of a cube, each of the sides must increase by a factor of the cube root of two, which is approximately 1.26.

$\sqrt{2}$ (1.4142)

The Pythagoreans worshipped whole numbers, believing they explained all the mysteries of the universe, from the harmonies in music to the beauties in nature to the movements of the stars.

It was a blow, then, when one of their own followers, Hippasus, found a number that could not be expressed as a ratio of whole numbers. This number was the square root of two, $\sqrt{2}$, and it can easily be drawn as the long side of a right-angled triangle whose two other sides have length 1. Yet it cannot be written as $^a/_b$ where a and b are whole numbers. Numbers expressible as $^a/_b$ are called rational, while those that are not are called irrational.

Hippasus met with an unhappy end for his blasphemous discovery. Today we know that there are infinitely many irrational numbers, and their infinity even outnumbers that of the rational numbers. The numerical value of $\sqrt{2}$ is 1.4142… – its decimal places never repeat, just as with any irrational number.

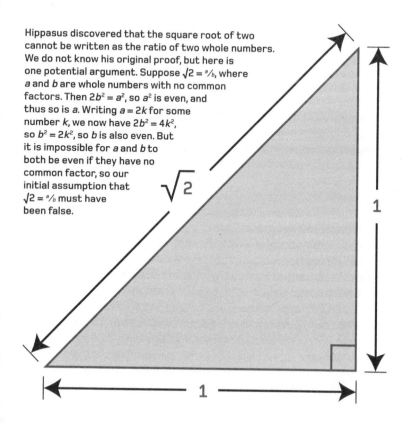

Hippasus discovered that the square root of two cannot be written as the ratio of two whole numbers. We do not know his original proof, but here is one potential argument. Suppose $\sqrt{2} = a/b$, where a and b are whole numbers with no common factors. Then $2b^2 = a^2$, so a^2 is even, and thus so is a. Writing $a = 2k$ for some number k, we now have $2b^2 = 4k^2$, so $b^2 = 2k^2$, so b is also even. But it is impossible for a and b to both be even if they have no common factor, so our initial assumption that $\sqrt{2} = a/b$ must have been false.

$\sqrt{2}$

1

1

1.4472

Five mathematicians are standing at the corners of a regular pentagon. Each one wishes to talk to their immediate clockwise neighbour, so off they set. However, the person they are pursuing is pursuing someone else, and so on. When will the mathematicians meet, and what path will they end up walking?

The answer is best summarized by the picture opposite. The mathematicians will walk in a spiral and will all meet in the centre of the pentagon. If the sides of the pentagon are one unit long, the length of the spirals are each 1.4472 units. More precisely, it is $(5 + \sqrt{5})/5$ km, or $1/(1 - \cos(72°))$. The precise shape of the spiral is a logarithmic spiral – the same as that seen on nautilus seashells, in the arms of galaxies or in the shape of cyclones. Curves of pursuit are used in biology to model animals pursuing prey, such as a hawk pursuing a mouse, though animals will often fly or run on a slightly less optimal curve in order to avoid being seen.

Pursuit curves

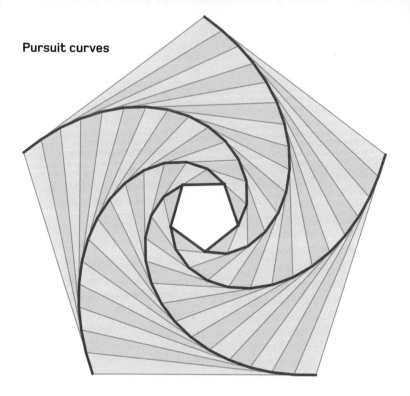

You can draw your own curves of pursuit by starting with a regular pentagon with sides of length 10 cm. Make new marks on the sides of the shape 1 cm clockwise from each corner. Join these to get a new pentagon. Repeat with the new pentagon! To get more precise curves, use marks 0.5 cm long, or smaller.

1.5236

Fold a strip of paper in half. Unfold it, and you have a V shape. Now fold it in half twice (in the same direction) before unfolding. The unfolded shape is now a little more complicated. Although you cannot fold a piece of paper in half more than about seven times, imagine doing this mathematically and consider the sequence of shapes that emerges.

You can analyse the shape of your paper by following the sequence of 90° turns and noting whether they are to the left or to the right. Whether you interpret the paper as a shape or as a string of Ls and Rs, the result is a fractal: an entity that is self-similar when you zoom in. The graphical interpretation is called a dragon curve. The outside edge of the dragon curve has a fractal dimension of about 1.5236, making it a shape in between a line (1D) and an area (2D). Yet the fractal dimension of the area inside the dragon curve is exactly two (see page 348). This makes it an example of a space-filling curve: a single line that fills out 2D space without any gaps. Nobody believed such a thing could exist until 1890.

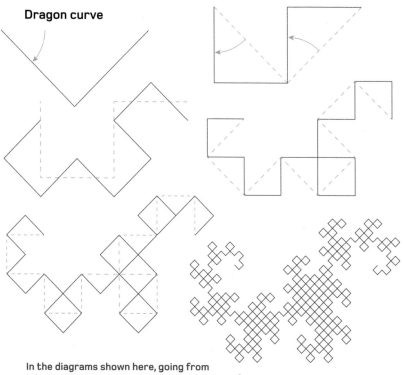

Dragon curve

In the diagrams shown here, going from
left to right in each one, we get the following
sequence of left (L) and right(R) turns:
L, LLR, LLRLLRR, LLRLLRRLLLRRLRR. Can you spot the pattern?

π/2 (1.5708)

Sometimes a random process can yield an accurate result. An excellent example of this is an experiment called Buffon's needle, in which you can calculate π using needles dropped at random onto ruled paper. To perform the experiment, start with a large collection of needles that are all the same length. On a sheet of paper, draw a series of parallel lines that are the same distance apart as the length of the needles. Drop the needles one by one onto the paper and count how many of them cross a line. The total number of needles divided by how many crossed a line is approximately π/2, or 1.5708. The approximation gets more accurate with the more needles that are dropped.

In fact, the needles do not have to be straight to achieve the same result! This leads to the amusing problem of Buffon's noodle, where the objects that are dropped are floppy noodles with the same length as the width between lines on the paper. In this experiment, however, you count all the times each noodle crosses a line, rather than a simple yes or no of whether it crosses a line at all.

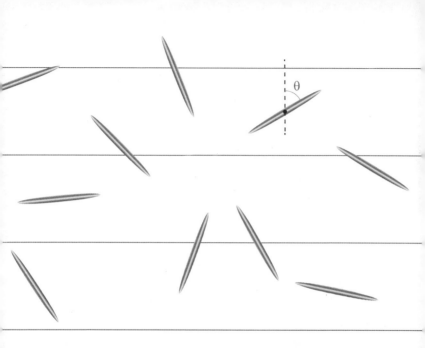

Ten needles have been dropped onto paper ruled with lines the same distance apart as the length of the needles. Six needles are crossing a line, giving an estimate of $^\pi/_2$ as $^{10}/_6 = 1.57$. The estimate improves as more needles are dropped. The reason π appears in this calculation is due to the trigonometry involved in the problem. If the distance from the centre of a needle to a line is less than $\cos(\theta)/2$ then the needle will cross a line. The exact probabilities can be worked out using some basic calculus.

1.585 $\left(\log3/\log2\right)$

With a fractal dimension of 1.585 (see page 348), the Sierpinski triangle appears unexpectedly in many different areas of mathematics. The basic construction is to start with an equilateral triangle, divide it into four smaller equilateral triangles and remove the centre one. Repeat the process with the remaining three triangles an infinite number of times. This produces a shape that is self-similar – that is, it looks the same however far you zoom in. Another way to produce the Sierpinski triangle uses randomness and is called the 'chaos game'. Once again, you start with an equilateral triangle. Draw a dot somewhere inside the triangle. Now pick one corner at random and draw a new dot halfway between the first dot and the chosen corner. Repeat the process with the new dot. You will need to draw a few hundred dots to see the pattern, but eventually the Sierpinski triangle will appear inside the triangle. A third method is to start with Pascal's triangle (see page 79) and colour in the odd numbers. Hence the Sierpinski triangle unites geometry, probability and numbers.

The Sierpinski triangle

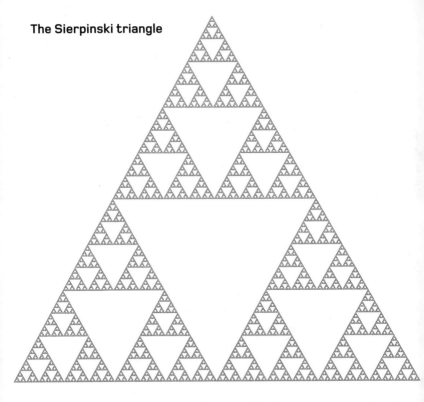

Having a fractal dimension of 1.585 (which is $\log(3)/\log(2)$) makes the Sierpinski triangle a shape that is somewhere in between a line and an area.

φ (1.6180)

The golden ratio (φ) is one of the most famous numbers in mathematics, and is equal to $(1 + \sqrt{5})/2$. It solves the puzzle of how to divide a line into lengths a and b so that the ratio of a to b is the same as the ratio of $a + b$ to a. When a and b are in the golden ratio, you can draw the golden rectangle, which can be divided into a square and a rectangle with the small rectangle having the same proportions as the large one.

Algebraically, the golden ratio solves the equation $x^2 - x - 1 = 0$. This results in some interesting properties. If you calculate $φ + 1 = 2.6180$, you get the square of φ, while if you subtract one to get $φ - 1 = 0.618$, you get its reciprocal $1/φ$. The golden ratio appears within the sequence of Fibonacci numbers 1, 1, 2, 3, 5, 8, 13,... where each number is the sum of the previous two (see page 50). Calculating the ratio of successive Fibonacci numbers gives the sequence of numbers $1/1$, $2/1$, $3/2$, $5/3$, $8/5$, ... etc, which evaluate to 1, 2, 1.5, 1.67, 1.6, ..., getting closer and closer to the golden ratio with each successive calculation.

The golden rectangle

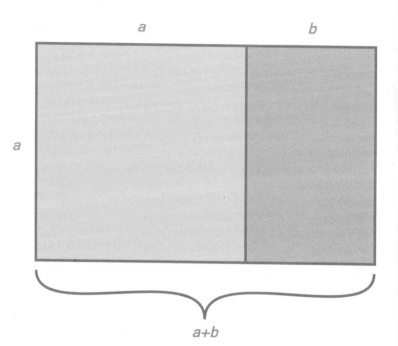

π²/6 (1.6449)

In 1650, an Italian named Pietro Mangoli posed a question about square numbers that had mathematicians bamboozled for nearly a hundred years. Three generations of the Bernoulli family tried and failed to solve it, but it became known as the Basel problem, for the town in which the Bernoullis lived. The problem asks for an exact answer to the sum of the reciprocals of the square numbers. In other words, it asks for a formula for the sum $\frac{1}{1} + \frac{1}{4} + \frac{1}{9} + \frac{1}{16} + \frac{1}{25} + \dots$, where the denominators are all the squares of the whole numbers. Although there are infinitely many terms in this sum, each number we add on takes the total closer and closer to a particular number, which is around 1.6449. That said, it requires adding more than one thousand terms before even the second decimal place is correct. It was another Basel resident, Leonhard Euler, who finally cracked the puzzle, coming up with an answer that nobody had expected: $\pi^2/6$. His insight was to connect the square numbers with the trigonometric sine function through a tool called the Taylor series.

$$\frac{\pi^2}{6} = \frac{1}{1^2} + \frac{1}{2^2} + \frac{1}{3^2} + \frac{1}{4^2} + \frac{1}{5^2} + ...$$

1.6829

The simple outline of the Mandelbrot set defies the complexity that surrounds it. The image opposite is a picture of the complex plane (see page 396), with real numbers plotted left-to-right and imaginary numbers (multiples of $\sqrt{-1}$) top-to-bottom. Each complex number is painted a different colour depending on how it behaves when it undergoes a particular sequence of squaring and adding. To find out the colour of a number c, you feed it into the formula $x^2 + c$ then feed the answer back into the formula, repeating to infinity. Numbers whose sequences are bounded – that is, they stay within a finite range – are coloured black, while those that are not bounded are coloured according to how quickly the sequence grows. The Mandelbrot set itself consists of all the complex numbers coloured black. Nobody yet knows what its exact area is. In 2015 a group of mathematicians calculated 1.6829 as a maximum possible area, though people counting high-resolution pixels put it closer 1.506. Given the incredible complexity of this object, it is possible that even counting trillions of pixels is not good enough for an accurate answer.

Mandelbrot set

The colour of a point c in the complex plane depends on how quickly the iterated function $x^2 + c$ goes to infinity. For example, when $c = 1$, you produce the sequence $1^2 + 1 = 2$, $2^2 + 1 = 5$, $5^2 + 1 = 26$, and so on, which grows quickly without bound. If you start with $c = 0.5$, however, you get $(-0.5)^2 - 0.5 = -0.25$, $(-0.25)^2 - 0.5 = -0.4375$, $(-0.4375)^2 - 0.5 = -0.30859375$, and so on, which stays bounded and is therefore coloured black.

√3 (1.7321)

In plane geometry, √3 appears in the most basic of shapes: the equilateral triangle, and hence in many common trigonometric formulae. In an equilateral triangle, where all the sides have length 1, the height of the triangle is $\sqrt{3}/2$ (a result derived from Pythagoras' theorem; see page 58).

This fact means that √3 is the diameter (the distance between a pair of parallel sides) of a regular hexagon where all the sides have length 1. It also means that sin(60°) is $\sqrt{3}/2$, and tan(60°) = √3. Less well known is that √3 also appears in the trigonometric calculations of 3°, 12°, 15°, 21°, 24°, 33°, 39°, 48°, 51°, 57°, 66°, 69°, 75°, 78°, 84° and 87°.

In three-dimensional geometry, √3 is the length of the long diagonal inside a cube with side lengths of 1. This diagonal is also known as a space diagonal. In general, for a cuboid of side lengths a, b and c, the space diagonal will have length $(a^2 + b^2 + c^2)$, in a generalization of Pythagoras' theorem.

Space diagonal

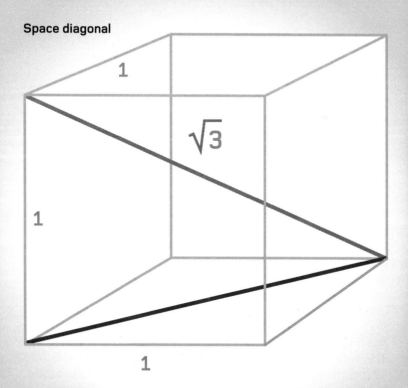

In a cube where the sides have length 1, the length of the long diagonal between opposite corners is $\sqrt{3}$. It is calculated from Pythagoras' theorem as $\sqrt{(1^2 + 1^2 + 1^2)}$.

$\sqrt{\pi}$ (1.7725)

What links the birth weight of babies, a blur filter on Instagram and the mathematical constants e and π? The answer is the Gaussian function. Imagine a plot of the various birth weights of babies. Many of them will be very close to average weight, a good number will be a little above or below the average, and a smaller number will be further away from the average. The shape of such a plot would be very close to the mathematical bell curve.

The mathematician Gauss worked out the formula for the shape of this graph. In its most general form, the height of the graph at a point x is e^{-x^2}, where e is the constant 2.718 (see page 346). The area below this graph is, surprisingly, $\sqrt{\pi}$. When statisticians use this graph, they must divide the formula by $\sqrt{\pi}$ to ensure that the total area (also the total probability) is equal to one. The Gaussian function e^{-x^2} is also used by graphic designers whenever they smooth a noisy image using Photoshop. Pixels further away from the centre of the blur are less affected than those in the middle, in the same proportions as the bell curve.

Gaussian function

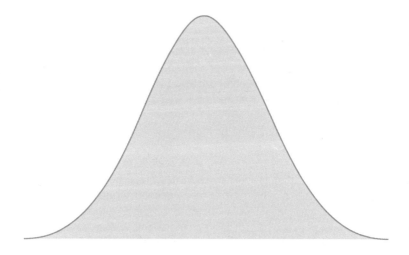

Statisticians refer to the Gaussian function as the bell curve or normal distribution.

1.9022 (Brun's constant)

Brun's constant is the sum of the reciprocals of the twin primes, and is approximately 1.9022. Twin primes are prime numbers separated by two. For example, 5 and 7, 17 and 19, 107 and 109. This is the closest separation that primes can have, since the even numbers above two are not prime. One of the big unsolved problems of number theory is whether there are infinitely many pairs of twin primes.

If you add up the reciprocals of all the prime numbers – that is, $1/p$ for all primes p, then the result is infinite, which itself is a proof that there are infinitely many primes. Viggo Brun used this method to attack the twin-prime conjecture in 1919, adding reciprocals of every pair of twin primes. That is, $1/p + 1/(p+2)$ where both p and $p + 2$ are prime. To his surprise, the result was a finite number, now called Brun's constant. The finite nature of Brun's constant does not prove that the twin primes are finite. However, if mathematicians can prove that it is an irrational number, this would solve the twin-prime conjecture.

$$\sum_{P,P+2 \text{ prime}} \left(\frac{1}{P} + \frac{1}{P+2} \right) =$$

$$\left(\frac{1}{3} + \frac{1}{5} \right) + \left(\frac{1}{5} + \frac{1}{7} \right) + \left(\frac{1}{11} + \frac{1}{13} \right) + \ldots$$

2.2195

If you have ever moved house and tried to manoeuvre furniture round a tight stairwell or corridor, you will identify with the unsolved mathematical problem of the moving sofa.

The sofa problem asks us to find the largest shape (as measured by its area) that can fit around an L-shaped corridor. The corridor in question has a width of one unit, so the height and width of the sofa can be, at most, one unit also. A semi-circular sofa of radius 1 will do the trick. This has an area of $\pi/2 = 1.57$, but it is possible to do a lot better than that. The sofa pictured opposite was designed by John Hammersley and consists of quarter-circles of radius 1 at each end, connected by a rectangle of length 4π with a semi-circle of radius $^2/_\pi$ removed. This sofa has an area of 2.2074, but it is still not the best we can do. The current best sofa, found by Joseph Gerver in 1992, has an area of 2.2195 and is made up of eighteen shapes patched together. Mathematicians believe that it is impossible to do better, but nobody has yet proven the result.

The sofa problem

What is the area of the biggest sofa that can fit round an L-shaped corridor?

2.4142

A lesser-known sibling of the golden ratio (see page 328), the silver ratio is equal in value to $1 + \sqrt{2}$. Like the golden ratio, it can be defined by ratios, by geometry or by algebra.

Two numbers a and b are in the silver ratio if the ratio of $2a + b$ to a is the same as the ratio of a to b. If you draw a rectangle that is in the silver ratio, you can remove two squares from the shape to leave a new rectangle of the same proportions as the original. There is also a connection with paper sizes, since if you take an A4 sheet of paper and cut off the largest possible square, the remaining rectangle will be in the silver ratio.

The silver ratio is important to geometers as it is the ratio of the lengths of a chord and a side of a regular octagon. Algebraically, the silver ratio is the number that solves the equation $x^2 - 2x - 1$. There is also a 'bronze mean' where the 2 in front of the x in this equation is replaced by a 3 and, indeed, a whole sequence of 'metallic means' for higher integers.

The silver ratio

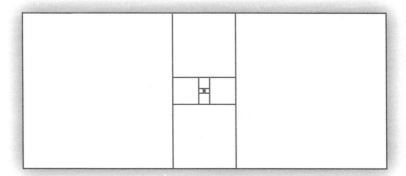

Cutting two of the largest squares possible from a silver rectangle leaves a silver rectangle on which the process can be repeated.

2.6651 (Gelfond–Schneider constant)

The square root of two, written $\sqrt{2}$, is an irrational number. That is, it cannot be written as a whole-number fraction. It is also an algebraic number, since it solves the polynomial equation $x^2 - 2 = 0$. If you raise two to the power of $\sqrt{2}$, you get a number called the Gelfond-Schneider constant, which is approximately 2.6651.

In 1900, German mathematician David Hilbert outlined twenty-three unsolved mathematical problems. Number seven was to show that a^b is always transcendental (that is, not algebraic) if a and b are algebraic and b is irrational ($a \neq 0, 1$). In 1919, he spoke of the transcendence of the Gelfond-Schneider constant as a problem of similar difficulty to the Riemann hypothesis (see page 378). However, it was proven true in 1930 by Rodion Kuzmin. Amazingly, an irrational to the power of an irrational can be rational. The square root of the Gelfond-Schneider constant can prove this even without knowing its own irrationality. If $\sqrt{2}^{\sqrt{2}}$ is rational then it proves the theorem. If it is irrational, then $(\sqrt{2}^{\sqrt{2}})^{\sqrt{2}} = (\sqrt{2})^2 = 2$, which is rational.

$$2\sqrt{2}$$

e (2.7182)

The mathematical constant *e*, approximately equal to 2.7182, is considered one of the most important numbers in mathematics, alongside 0, 1, π and *i*. It is fundamental to calculus, is the base of the natural logarithm and is important in the study of compound interest.

Appearing gradually, in different works of the seventeenth century, *e* is a relative mathematical newcomer. It was Euler who first used the notation *e* in 1731. In calculus, e^x is the unique curve whose rate of change is equal to itself. For this reason, it is called 'the' exponential function, rather than 'an' exponential function. If $y = e^x$ then $x = \ln y$, the natural logarithm of *y*, which is very important in science (see page 296).

The decimal expansion of *e* is infinite and non-repeating, as *e* is an irrational number. It can be calculated by adding the reciprocals of the factorial numbers: $\frac{1}{1!} + \frac{1}{2!} + \frac{1}{3!} + \ldots$ (Here *n*! means the product of all the numbers from 1 to *n*, for example, $3! = 3 \times 2 \times 1$.)

Compound interest

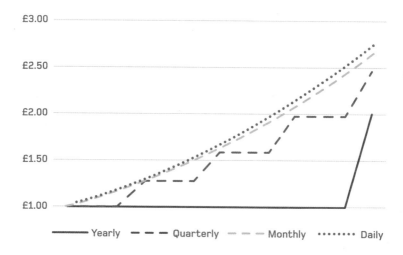

Would you rather receive 100% interest once per year, or 50% interest twice per year? How about 25% interest four times per year? Your instinct might be that all these options provide the same return, but the options are becoming progressively more lucrative because of compound interest. If you invested £1, 100% interest after a year would give you £2. But 50% twice per year gives you $£1 \times (1.5)^2 = £2.25$, and 25% four times per year gives you $£1 \times (1.25)^4 = £2.44$. The mathematical constant e is the maximum return if interest were being added and compounded continuously throughout the year.

2.7268

The Menger sponge has infinite surface area but zero volume. Appearing self-similar on every scale, it is a fractal. To build the Menger sponge, divide a cube into twenty-seven smaller cubes in a $3 \times 3 \times 3$ arrangement. Then remove the middle cubes on each of the six sides as well as the centre cube, leaving twenty small cubes. Now repeat this dividing and removing procedure on each of the small cubes, iterating to infinity. Each time you perform a divide-and-remove procedure, creating the next level of the sponge, the volume is multiplied by $^{20}/_{27}$. Performing this multiplication repeatedly causes the volume to get closer and closer to zero. You are then left with a shape that has area but no volume, so is it two-dimensional or three-dimensional? The concept of fractal dimension can help here. The idea is that, if you scale your shape by a factor of $\frac{1}{3}$, you should be able to fit 3^d copies of it into the old shape, where d is the dimension. If the Menger sponge is $\frac{1}{3}$ the size, you can fit twenty copies of it into the big one, making the dimension 2.727 (as $3^{2.727} = 20$). Thus the sponge is a shape in between two and three dimensions.

Menger sponge

A different way of building a Menger sponge is to build up instead of dividing down. A Level 0 sponge is a single cube. A Level 1 sponge has twenty cubes arranged (as shown) in a larger cube. A Level 2 cube has twenty Level 1 cubes arranged in the same way, and a Level 3 consists of twenty Level 2 cubes.

π (3.1415)

Ask people to think of a famous mathematical number, and *pi* (π) will probably come top of the list. Few numbers have inspired so much in our culture and media, from songs and films to memory feats and giant calculations. *Pi* is the ratio between the circumference of a circle and its diameter. No matter how big a circle you draw, this ratio is always the same, equal to about 3.14. The true answer is an irrational number, with a decimal expansion whose digits never repeat. The digits of π have so far passed all tests for true randomness. It is conjectured that π is a normal number, meaning that any string of digits should appear in π with the same frequency as any other string of the same length. For example, my birthdate or phone number should appear as often as yours. This randomness makes memorizing the digits of π a feat worth celebrating: the current record is 70,000 digits, by Rajveer Meena in India. In 2019 a new record was set for computing the digits of π, with Emma Haruka Iwao finding 31.4 trillion decimal places. Yet only thirty-nine digits are needed to compute the circumference of the universe to the precision of a single atom.

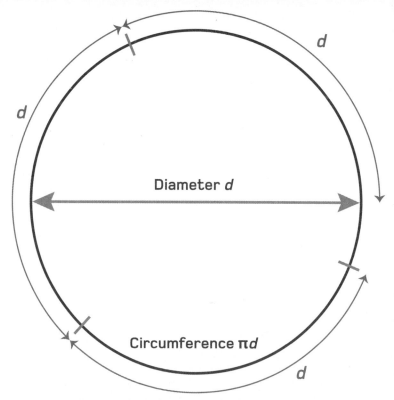

Diameter _d_

Circumference π_d_

A little over three diameters can fit around the circumference of a circle.
The exact number is an irrational number called π, equal to 3.14159...
The area of a circle of radius 1 is also equal to π.

$^{22}/_7$ (3.1429)

Maths lovers enjoy celebrating on *Pi* Day: March 14 (3.14). Yet the true π fans will be found celebrating on 22 July, since $^{22}/_7$ is a much better approximation for π than 3.14. We know that *pi* is an irrational number (see page 264). This means it cannot be written as $^a/_b$ with *a* and *b* whole numbers. Yet fractions (rational numbers) are far easier to work with, so people throughout history have attempted to find good rational approximations to π. So, what makes for a good rational approximation? Accuracy and simplicity. For example, $^{22}/_7$ and $^{179}/_{57}$ are both accurate to two decimal places, but $^{22}/_7$ is considered a better approximation because of its smaller denominator. Another notably good fraction for π is $^{355}/_{113}$, which is accurate to six decimal places.

Historians disagree about who first discovered $^{22}/_7$ as an estimate for π, but we know that Archimedes knew of it by the third century BCE. Using ninety-six-sided polygons to approximate a circle, he showed that π lay between $^{223}/_{71}$ and $^{22}/_7$. Averaging these numbers captures π to within an $^8/_{1000}$th of a per cent.

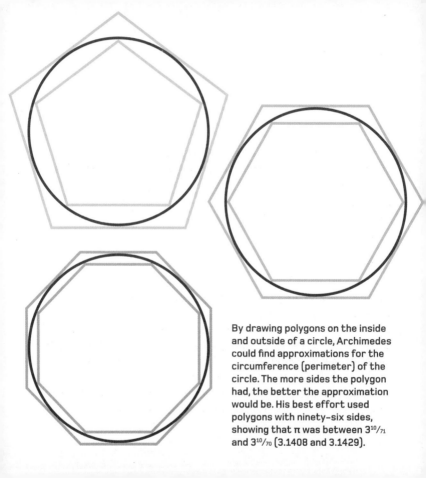

By drawing polygons on the inside and outside of a circle, Archimedes could find approximations for the circumference (perimeter) of the circle. The more sides the polygon had, the better the approximation would be. His best effort used polygons with ninety-six sides, showing that π was between $3^{10}/_{71}$ and $3^{10}/_{70}$ (3.1408 and 3.1429).

3.57

Simplicity does not imply predictability. This is the conclusion we can draw from studying the logistic map: a very simple model of population growth that rapidly gives chaotic results.

To model a population of animals, you want an equation that grows quickly when the number of animals is small (no competition for resources) but slows down as numbers reach the maximum that the environment can hold. The logistic map is the simplest equation satisfying these requirements. If the symbol x_n means the size of the population at time n, then the logistic map is $x_{n+1} = rx_n(1-x_n)$. Here r is a parameter relating to the birth rate. As we vary r, the population's behaviour changes. For r less than 1, the population dies out since they are not having enough babies. When r is between 1 and 3 the population reaches a stable size. When r is between 3 and 3.45 the population fluctuates between two different sizes, then four then eight. Suddenly, at the magic number of 3.57, the population fluctuates wildly with no discernible pattern. This is called the onset of chaos.

The onset of chaos

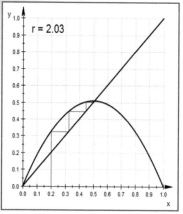

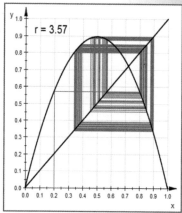

The logistic equation is a simple quadratic equation that models population growth. The variable r reflects the birth rate, and it dramatically influences behaviour. When r is around 2, on the left, the population approaches a stable size. Yet when r is 3.57 and above (right), the population size varies wildly with no pattern or apparent predictability.

3.708

You would be forgiven for looking at a squircle and thinking it was just a rounded square. But it is not, and product designers are catching on to this. A rounded square is a square with its corners replaced with arcs of quarter-circles. The shape is conceptually straightforward but has two problems for designers: its mathematical equation is complicated, and its curvature does not vary smoothly. In particular, the curvature suddenly changes at the points where the quarter-circles join to the straight edges.

A squircle, however, has the simple equation of $x^4 + y^4 = r^4$. This produces a shape very similar to a rounded square of side lengths $2r$. The difference is that, in the squircle, the curvature varies smoothly all the way around the shape and so has a slightly smaller area of 3.708 if the radius is 1. Squircles are most commonly spotted in mobile phone designs. Nokia have used the shape for touchpad buttons in many of their phones, while the Android operating system (as of 2017) allows users to choose squircular icons.

A Squircle

A squircle (dark grey) overlaid on a rounded square (light grey). Can you see the difference? A squircle with radius 1 has an area of 3.708: less than a square (area 4) but more than a circle (area 3.14).

3.7224

H ave you ever wondered why bees build their honeycombs in a hexagonal pattern? Why not use squares or triangles, or even completely irregular shapes? Writing in around 300 CE, Pappus of Alexandria tried to answer the question: 'Bees... know...that the hexagon...will hold more honey for the same expenditure of material [than the square or triangle].' The honeycomb conjecture, as it is known today, states that, out of all the different ways of dividing a plane into shapes of equal area, the hexagonal honeycomb is the one that has the least perimeter. The perimeter of a hexagon with area 1 is $2(\sqrt[4]{12})$ = 3.7224. In the infinite tiling each edge is shared between two tiles so $\sqrt[4]{12}$ is the average perimeter per tile. As obvious as this was to ancient scholars, it took until 1999 before mathematicians had a proof. Polygons with more sides, such as the octagon and decagon, have a better ratio of perimeter to area, but the advantage is offset by the fact that they do not tile the plane (tessellate to fill the space without gaps). So, out of all straight-sided shapes, regular hexagons are the best.

The honeycomb conjecture

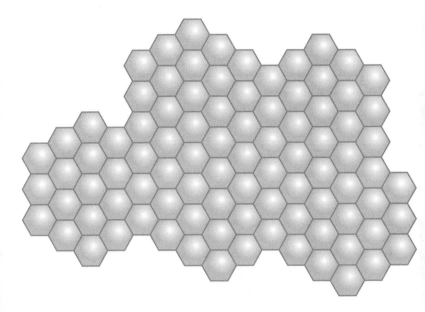

As hexagons are placed together in a tiling, without gaps, they cover the plane in such a way that the total perimeter is the smallest among all tilings of shapes with the same area.

4π/3 (4.1887)

Before he died, the ancient Greek mathematician and scientist Archimedes asked that his tomb be inscribed with a picture of a sphere inside a cylinder. It represented one of the results that Archimedes was most proud of proving: that the volume of a sphere was two-thirds the volume of the enclosing cylinder. Or, in other words, that the volume of the sphere was twice the volume in between the sphere and the enclosing cylinder.

Doing the calculation, this means that the volume of a sphere of radius r is $4\pi r^3/3$. If the radius is equal to 1, this gives the volume as about 4.19. It is slightly more than half of the enclosing cube, which has a volume of eight.

Today we would derive this result using integral calculus, but Archimedes would not have had this tool in his canon. Based on recovered works of his, it seems that he worked it out using balancing scales and a clever argument about centres of mass.

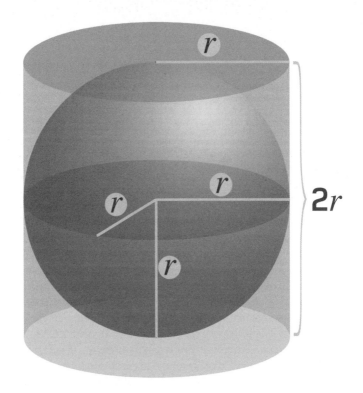

For a sphere of radius r, the volume of the enclosing cylinder is the product of the base (πr^2) and the height ($2r$), giving $2\pi r^3$. The volume of the sphere is two-thirds of this.

8π²/15 (5.2638)

It is a curious fact that spheres in five dimensions contain more volume than spheres in any other dimension. What does it mean to have a sphere in a dimension higher than three? A sphere of radius r is defined as the set of all points a distance r away from a fixed point (the centre of the sphere). In two dimensions this is a circle, and in three dimensions it is the ball shape we are familiar with. In higher dimensions, though we can no longer draw pictures, we can analyse the situation using coordinates and algebra. For example, in four-dimensional space, each point has four coordinates: (x, y, z, w). A sphere of radius 1, centred on 0, is all those points with $\sqrt{(x^2 + y^2 + z^2 + w^2)} = 1$. The formula for the volume of a sphere of radius 1 is different for even and odd dimensional space. In dimension $2k$, it is $\pi^k/k!$ (where $k!$ means the product of all the numbers up to k), while in dimension $2k+1$ it is $2^{k+1} \pi^k/(2k+1)!!$ (where $(2k+1)!!$ means the product of all the odd numbers up to $2k+1$). Doing the calculations gives the volume in five dimensions as $8\pi^2/15$. As the dimension increases after that, the volume of the spheres gets closer and closer to zero.

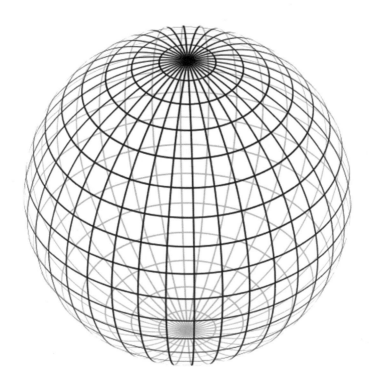

A sphere in three dimensions.

$\pi + e$ (5.8599)

Numbers can be divided into two categories: rational and irrational. The former can be expressed as whole-number fractions, such as $5/3$ or $3/11$, while the latter cannot. Rational numbers always have a decimal expansion that is either finite or repeats a finite sequence of digits. For example, $3/11$ equals 0.272727... with the 27 repeating. Most numbers are irrational. A decimal picked at random will almost certainly not have repeating digits. Yet, despite the lack of pattern to their decimals, some irrationals can easily be described using algebra. The square root of two is one example: it solves the equation $x^2 - 2 = 0$. Numbers that solve these kinds of equations (using only powers of x and whole numbers) are called algebraic numbers. Most numbers are not algebraic. Those that are not are called transcendental. The famous constants π and e are both transcendental, yet very few transcendental numbers are known. It is still an open question whether $\pi + e$ is transcendental, alongside $\pi - e$, $\pi \times e$, π/e and π^e. However, we do know that e^π is transcendental.

$$\pi + e$$

$$\pi - e$$

$$\pi \times e$$

$$\pi / e$$

$$\pi^e$$

Transcendental numbers are the most common numbers of all, yet very few are known. Although π and *e* are both transcendental, it is unknown whether any of these combinations are also transcendental.

2π (6.2832)

We are used to measuring angles in degrees, so that there are 360° in a circle. But there is something unsatisfying about this to a mathematician: 360 is an arbitrary number that has nothing to do with circles. Wouldn't it be better to have a measure of angle that is somehow intrinsic to the geometry of a circle?

These musings led to the introduction of the radian, now the SI unit (international standard unit) for angle. In a circle of radius r, turning through an angle of 1 radian traces out an arc of length r along the circle. If your circle has a radius of one unit, this means that the arc length around the circle is the same as the number of radians it subtends – that is, the angle it draws out at the centre of the circle. How many radians are in a circle? The answer is 2π (or about 6.28), since 2π radiuses fit around the circumference of a circle. This means that 360° equals 2π radians, and one radian is $180/\pi = 57.3°$ degrees.

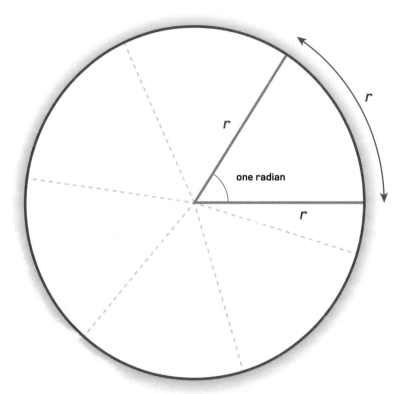

An angle of one radian traces out an arc of a circle of length equal to the radius of the circle. Since the circumference of the circle is $2\pi r$, this means there are 2π, or just over six, radians in a circle.

$^{41}/_{6}$ (6.8333)

In the thirteenth century, Leonardo of Pisa (also known as Fibonacci) was presented to the court of the Holy Roman Emperor Frederic II. The emperor's scholars challenged him to find a right-angled triangle with rational number sides (ratios of whole numbers) and an area of 5. Leonardo found the solution: the sides have lengths $^{3}/_{2}$, $^{20}/_{3}$ and $^{41}/_{6}$. But Leonardo wondered whether it was possible to find rational-sided right-angled triangles with any whole-number area. It was a question that had been asked 200 years earlier in an Arab manuscript.

The answer is that not every whole number can be the area of such a triangle. Fermat (400 years after Fibonacci) showed that it was impossible for 1, 2, 3 and 4. Numbers that it does work for are called congruent numbers. In 1983, mathematician Jerrold B. Tunnell gave a simple test to check whether a given number was congruent. Yet its proof relies on the Birch and Swinnerton-Dyer conjecture – a Millennium Prize problem worth one million dollars. So Fibonacci's simple triangle problem is yet awaiting a solution.

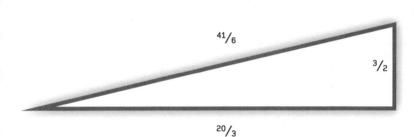

$^{41}/_6$

$^3/_2$

$^{20}/_3$

The area of this right-angled triangle is 5.
Can you find triangles with rational side
lengths whose areas are 6 or 7?

3π (9.4248)

When you throw a ball through the air from one person to another, it follows the shape of a parabola (or $y = x^2$ shape). But what if you want to drop the ball down a slide to the waiting catcher? What shape should the slide be to get the fastest descent? Galileo had tried to solve the problem in 1638, naming an arc of a circle as the answer, but he was incorrect.

In 1696 Johann Bernoulli set a challenge to find a solution within two years. Five mathematicians rose to the challenge, including Isaac Newton, who claimed he only heard of the contest the day before the deadline, and stayed up all night to solve it. The answer to the problem is a curve called a cycloid. The cycloid is also the shape traced out by a point on a wheel as it rolls along a straight line. The name cycloid was coined by Galileo, but he did not have the mathematical tools to realize it solved his problem. However, he did figure out the area under the cycloid: if the wheel has radius 1 the area traced out is 3π (9.4248). Cycloids have been used in architecture and pendulums, as well as to design skating ramps.

The cycloid

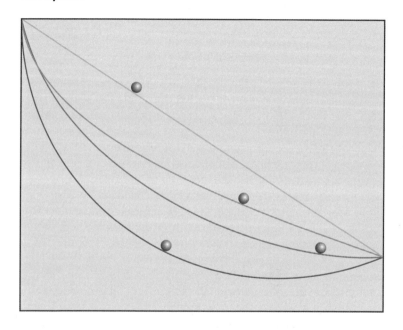

If a ball were dropped from the top left of the image, which shape ramp would make it slide to the bottom first? The straight line (top) is the slowest, followed by the circle (bottom) and parabola (second from top), with the cycloid (third from top) being the winner.

π^2 (9.8696)

A s you sit reading this book, Earth's gravity is pulling you towards it with a force of g = 9.81 Newtons per kilogram (that's an acceleration of 9.81 metres per second per second). Curiously, this number is very close to the value of π^2, or $\pi \times \pi$, which is 9.8696. The two numbers, in fact, have a close history, connected to our definition of the metre. For most of human history there was no standard measure of length. In the seventeenth century there were calls for the creation of a single standard that would be based on a natural phenomenon – but what to pick? English polymath Christopher Wren suggested that one metre should be the length of a pendulum that would take one second to swing from one side to the other. This would have defined the metre as g/π^2. The problem with this was that g varied from place to place on the Earth (in Kuala Lumpur, for example, it is only 9.776 ms^{-2}). In the end, one metre was defined as one ten-millionth the distance from the North Pole to the equator along the meridian passing through Paris. This made the seconds pendulum only 0.994 m long and π^2 not quite equal to g.

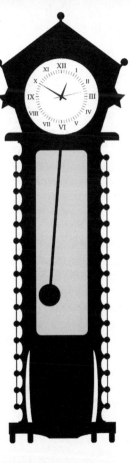

The time a pendulum takes to swing from one side to the other does not depend on how far out you hold it, or how heavy a weight is at the bottom, but only on its length. A pendulum of length 0.994 m will take one second in its swing.

10.47

When you look at a map of the world, countries like Alaska, Greenland and Antarctica look very big. But don't be fooled: you are probably looking at a Mercator projection, in which case those countries are much smaller than they seem. It is mathematically impossible to draw an accurate map of the Earth, since the Earth is a sphere and a map is flat. Different map-makers choose to preserve different features: some preserve distance, others area or shape, but all will distort something. The Mercator projection, designed in 1569 by Gerardus Mercator, is still commonly used today for navigation. It involves a cylindrical projection: imagine wrapping a cylinder around the equator, projecting and then unrolling. This preserves angles but distorts distance, especially in countries far from the equator. For Greenland, at an average latitude of 72°N, the area distortion is $(1/\cos(72))2 = 10.47$, meaning that areas appear about ten times bigger than they really are. This results in Greenland appearing to be almost as big as Africa, despite actually being smaller than Algeria.

Mercator projection

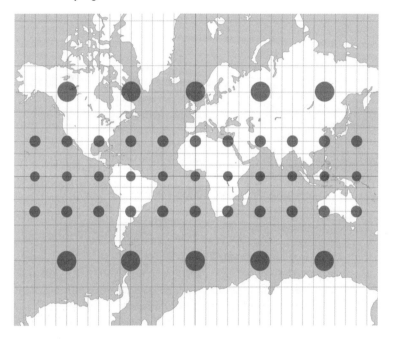

On a Mercator projection of the world, countries near the North and South Poles look much bigger than they really are. The exact variation in scale is indicated by the circles on this map. On a spherical globe, all these circles would be the same size.

4π (12.566)

The number 4π (12.566) appears in geometry in at least two different contexts. The first is that 4π is the surface area of a sphere of radius 1. This is because the formula for the surface area of a sphere of radius r is $4\pi r^2$. Archimedes was the first to record this fact, and he found it by realizing that the answer was the same as the surface area of a cylinder enclosing the sphere (excluding its top and bottom). This is its height ($2r$) multiplied by its circumference ($2\pi r$), giving $4\pi r^2$.

The second appearance of 4π is in the field of knot theory, where it marks the lower bound of how much curvature a knot can have. A mathematical knot is a curve in three-dimensional space, where the end of the curve must join to the beginning in an infinite loop. Sometimes a curve may seem very complicated, but if we make it out of string and pull it about cleverly, it untangles into a simple loop. Such knots are called unknots. The Fary-Milnor theorem says that if your loop's total curvature is less than 4π then it must be an unknot.

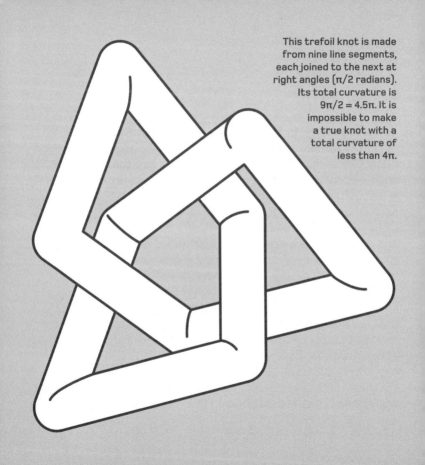

This trefoil knot is made from nine line segments, each joined to the next at right angles ($\pi/2$ radians). Its total curvature is $9\pi/2 = 4.5\pi$. It is impossible to make a true knot with a total curvature of less than 4π.

14.1347

The Riemann hypothesis concerns the distribution of the prime numbers, suggesting that their placement is governed by certain imaginary numbers. At the centre of the hypothesis is an object called the Riemann zeta function. It takes in a number, s, and outputs the sum of the reciprocals of the whole numbers, each to the power of s: $\frac{1}{1^s} + \frac{1}{2^s} + \frac{1}{3^s} + \frac{1}{4^s} + \frac{1}{5^s} + \cdots$

This expression is well understood when s is a real number bigger than one, but it becomes a different beast when s is allowed to be a complex number – that is, a mixture of real and imaginary numbers, such as $2 + 3i$ (where $i = \sqrt{-1}$). Certain complex numbers cause the Riemann zeta function to output zero. The 'trivial zeros' are the negative even numbers; the 'non-trivial zeros' are all the rest, and these are what govern the placement of the primes. The smallest example of a non-trivial zero is $\frac{1}{2} + 14.135i$. The Riemann hypothesis is that all such non-trivial zeros start with $\frac{1}{2}$. Over ten trillion zeros have been checked and the hypothesis holds, but who knows if the next discovered zero will break the result?

The Riemann hypothesis

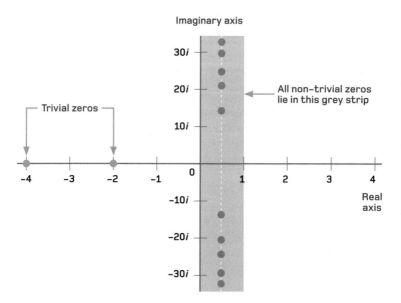

The Riemann zeta function has two different collections of zeros, marked as grey dots. The trivial zeros lie on the horizontal axis on each of the negative even numbers. The non-trivial zeros lie in the grey strip. More than that: the non-trivial zeros seem to lie along a single straight line. Whether or not this is true of *all* the non-trivial zeros is one of several Millennium Prize problems, worth one million dollars to the person who solves it.

22.92

When a string vibrates with a wavelength that is $\frac{1}{2}$, $\frac{1}{3}$, $\frac{1}{4}$ … of its fundamental wavelength, the notes sound especially harmonious together. This inspired the name of the harmonic series, which is the sum of the reciprocals of the whole numbers: $1 + \frac{1}{2} + \frac{1}{3} + \frac{1}{4} +$… This series is an example of a sum where the terms grow smaller and smaller, yet whose total sum is infinite. This contrasts with infinite sums such as $\frac{1}{2} + \frac{1}{4} + \frac{1}{8} +$… which get closer and closer to a particular number. Yet the harmonic series is only *just* infinite. If you remove all the numbers whose denominators include a 9, the remaining sum adds up to 22.92. In fact, removing any terms whose denominators contain a particular string of digits (for example, 193610361863) causes the harmonic series to have a finite sum. The harmonic series leads to counterintuitive results. For example, imagine a worm crawling along a rubber band at 1 cm per minute. After each minute the band is stretched by 1 m. Will the worm ever reach the end of the band? It seems impossible but, using the harmonic series, it turns out the answer is 'yes' (though after more minutes than the age of the universe!).

$$\frac{1}{1} + \frac{1}{2} + \frac{1}{3} + \frac{1}{4} + \frac{1}{5} + \frac{1}{6} + \ldots = ?$$

109.47

A tetrahedron is the simplest of the five Platonic solids – highly symmetric shapes whose faces are all regular polygons. It is built from four equilateral triangles, and is a special kind of triangular-based pyramid. Tetrahedral geometry is important in chemistry, as a number of important molecules have their atoms arranged at the corners of tetrahedra. For example, methane molecules (CH_4) and ammonium ions (NH_4^+) have four hydrogen atoms arranged tetrahedrally around a central carbon or nitrogen atom respectively. The tetrahedral angle, which is the central angle between two of the vertices, is approximately 109.5°. In a methane molecule, this would be called the bond angle. This angle is very close to the angle between two hydrogen atoms in a water molecule, which is 104.5°. The two hydrogen atoms and two lone electron pairs in H_2O form a tetrahedral structure, but the electron pairs repel each other slightly, pushing the hydrogens closer together. This slight asymmetry in the tetrahedral structure is what gives water many of its important properties. Without it, water would not be a liquid, and we would not be alive!

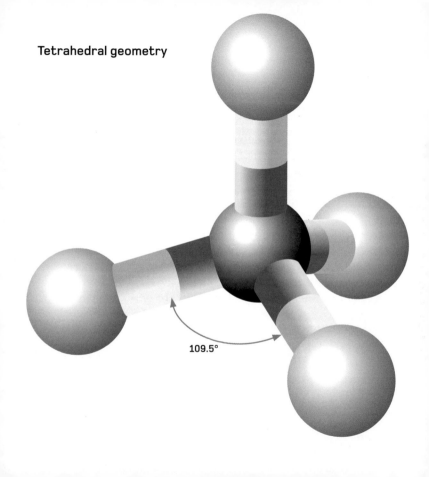

Tetrahedral geometry

109.5°

111.11

In a paradox known as 'Achilles and the tortoise', the ancient Greek philosopher Zeno of Elea shows that it is impossible for a faster runner to overtake a slower one. Suppose that, each minute, Achilles can run 100 m, while the tortoise can only crawl 10 m. Achilles gives the tortoise a head start of 100 m to make the race fair. After the first minute, Achilles has run 100 m and has caught up to where the tortoise started. But the tortoise is now 10 m ahead. So Achilles runs 10 m to catch the tortoise, but the tortoise has travelled a further metre. When Achilles runs this extra metre, the tortoise has travelled 0.1 m. It seems that whenever Achilles tries to catch the tortoise, the tortoise is always just ahead. In modern language, Zeno is asking us to calculate the infinite sum $100 + 10 + 1 + 0.1 + 0.01 + 0.001 + \ldots$ to find the point where Achilles catches the tortoise. Although it is 'obvious' that this point exists, it took until the invention of calculus to truly resolve the paradox. Indeed, Achilles will overtake the tortoise at a distance of $111.11 = 1000/9$ m.

Zeno's paradox

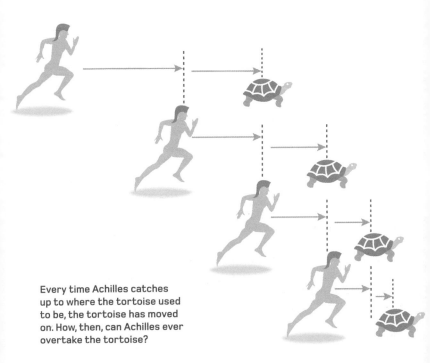

Every time Achilles catches up to where the tortoise used to be, the tortoise has moved on. How, then, can Achilles ever overtake the tortoise?

137.5

If you divide the 360° of a circle into the golden ratio (see page 328), you get the two angles of 222.5° and 137.5°. The smaller of the two is called the golden angle.

The golden angle is important in phyllotaxis, which is the study of how leaves arrange themselves on plant stems. Suppose that a plant arranged each leaf at right angles, or 90°, to the one beneath. This would be fine for the first four leaves, but then the first one would block the light from the fifth one. It turns out that the golden angle is the best angle to use so that each leaf has the most sunlight without blocking the others.

If leaves arrange themselves according to the golden angle, then they will form spirals where each spiral has a Fibonacci number of leaves, due to the relationship between Fibonacci numbers and the golden ratio. This also occurs in petals and seed heads, which explains why sunflower heads contain Fibonacci spirals.

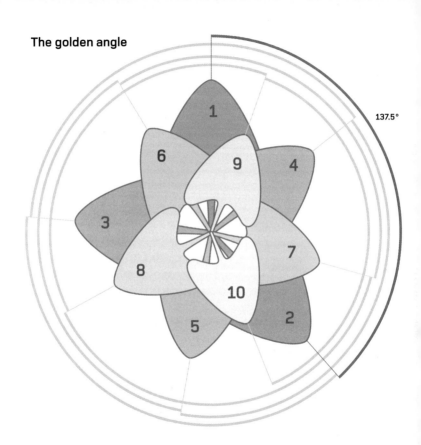

The golden angle

137.5°

$-1/12$

It is possible to show, in a mathematically rigorous and correct way, that $-1/12$ is the sum of the natural numbers. That is, $1 + 2 + 3 + 4 + 5 + \ldots = -1/12$. Yet this is clearly an absurd conclusion, since summing positive whole numbers cannot possibly result in a negative fraction.

Many incorrect 'proofs' of the result can be found on the internet, using manipulations of the infinite sum that are self-contradictory. That is, similar manipulations of the same kind would yield different answers. The correct way of arriving at the answer is to use a complicated tool called the 'analytical continuation of the Riemann zeta function'. This uses the heavy machinery of complex numbers and calculus to arrive at a consistent result.

Absurd and esoteric though the result may seem, it has been used by physicists in string theory. It is what ultimately leads to the result that we may live in a twenty-six-dimensional universe.

The Indian mathematician Srinivasa Ramanujan was a true mathematical genius with a remarkable intuition for numbers. An extract from a 1919 notebook shows that the sum of $1 + 2 + 3 + 4 + \ldots$ is equal to $-1/12$. But his manipulations do not hold up to the rigorous scrutiny of modern mathematicians. In particular, we cannot add multiples of infinite sums together in this way without potential contradictions.

Another way of finding the constant is as follows — 41.
Let us take the series $1 + 2 + 3 + 4 + 5 + \&c$. Let C be its constant. Then
$$c = 1 + 2 + 3 + 4 + \&c$$
$$\therefore 4c = \quad 4 \quad + 8 \quad + \&c$$
$$\therefore -3c = 1 - 2 + 3 - 4 + \&c = \frac{1}{(1+1)^2} = \frac{1}{4}$$
$$\therefore c = -\frac{1}{12}.$$

−1

We use negative numbers every day: a negative number on the Celsius temperature gauge means it will be below freezing; a negative number in our accounts means we owe money to somebody; a negative number in an elevator means the floor is below ground level. Yet negative numbers are relatively young in mathematical history, and were avoided as recently as the eighteenth century among Western mathematicians.

Negative numbers – those that are less than zero – can be thought of as opposites of positive numbers. Together, positive and negative whole numbers make up the integers. Adding a negative number to its positive twin always gives an answer of zero: $(-1) + (1) = 0$. A negative number can also be thought of as a subtraction, so −1 is the result of taking 1 away from 0. The seventh-century Indian mathematician Brahmagupta was the first known person to write down the rules of arithmetic for negative numbers. For example, adding two negatives gives a negative number, while multiplying two negatives gives a positive number.

The number line

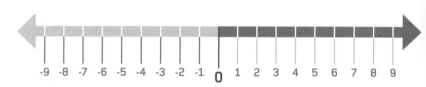

On a traditional number line, the positive numbers are drawn to the right of zero, getting ever larger, with negative numbers extending to the left, getting ever smaller. For example, 5 is larger than 2, but −5 is smaller than −2. Addition means counting to the right, while subtraction means counting to the left. For example, 3 − 5 means starting from 3 and moving 5 units to the left, giving an answer of −2.

−5

Mathematicians consider prime numbers the building blocks of all other numbers. This is because every whole number can be written uniquely as a product of prime numbers. For example, 12 can be factorized as (1 × 12), (2 × 6) or (3 × 4), but in terms of primes it can only be written as (2 × 2 × 3). In fact, we are very lucky to work with a number system in which prime factorization is unique like this. When mathematicians started looking more widely into different number systems based on the integers, they found that it was a rare property indeed. Let us look at what happens if we take the number system that is the integers with $\sqrt{-5}$ included as an extra number. This means we can add, subtract and multiply numbers as usual, but we can also add, subtract and multiply $\sqrt{-5}$ as well. Then it turns out 6 has two different 'prime' factorizations: (2 × 3) and $(1 + \sqrt{-5})(1 - \sqrt{-5})$. Here $(1 \pm \sqrt{-5})$ are 'prime' in the sense that they cannot be written as products of two smaller numbers. The discovery of such number systems has led to deep and profound research into numbers, and a questioning of what 'prime' really means.

Emmy Noether, born in Germany in 1882, made groundbreaking progress in the study of different number systems and the ideas of prime factorization. Her achievements are all the more impressive for being made in a hostile world, where as a female student she was not permitted to fully attend lectures, and as a female lecturer she had to work without pay. Later in her life she was forced out of her university with the rise of the Nazis.

$i\,(\sqrt{-1})$

The square root of –1, denoted by the symbol i, is called an imaginary number because it doesn't exist anywhere on the real number line. Two positive numbers multiplied together give a positive number, and similarly two negative numbers multiplied together give a positive number, so a real number cannot be multiplied by itself to give –1. Graphically, i is drawn on a new axis perpendicular to the real numbers. This is called the imaginary axis. The square root of any other negative number can be expressed as a multiple of i (e.g. $\sqrt{-4} = 2\sqrt{-1} = 2\,i$), so all imaginary numbers lie on this axis. Combinations of real and imaginary numbers, such as $2 + 3i$, are called complex numbers and are drawn in the plane using the real and imaginary axes as coordinate systems. Complex numbers provide a two-dimensional number system, which makes them highly useful in areas of physics and engineering. For example, quantum physics uses complex numbers to keep track of both position and momentum of particles, while in signal processing complex numbers can track the magnitude and phase of waves.

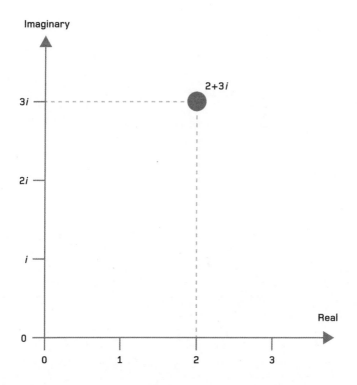

$$-\frac{1}{2} + \frac{\sqrt{3}}{2}i$$

W hich number multiplies by itself to give 1? Pat yourself on the back if you not only answered '1', but also '–1'. Indeed, every number has two square roots: one positive and one negative. Now a harder question: what number multiplied three times gives 1? (That is, what is the cube root of 1?) Clearly 1 is again an answer, since $1 \times 1 \times 1 = 1$. It is NOT the case that –1 is a root, since three negatives multiplied together would give a negative.

Just as square roots always have two answers, so it turns out that cube roots always have three answers, except the other two are hidden away among the complex numbers: those involving i, the square root of –1. Crunching the numbers, we discover that $-\frac{1}{2} + \frac{\sqrt{3}}{2}i$ is also a cube root of one, as is $-\frac{1}{2} - \frac{\sqrt{3}}{2}i$. Plotting these numbers in the complex plane, together with our first root, 1, we find a wonderful connection between arithmetic and geometry. The three roots are equally spaced around a circle. This discovery is no coincidence. There are always n different nth roots of a number, equally spaced around a circle in the complex plane.

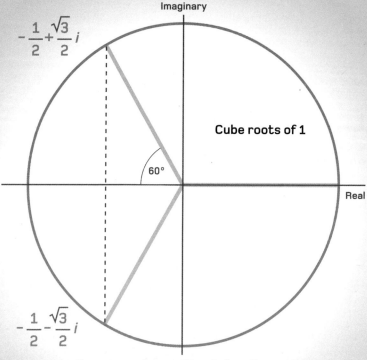

The cube roots of one are equally spaced in a circle on the complex plane. You can find the roots by using some basic trigonometry. The root on the top part of the circle forms a right-angled triangle with a hypotenuse of 1 and an angle of 60° with the horizontal axis. This gives the horizontal (real) distance as $1/2$ in the negative direction, and vertical (imaginary) distance as $\sqrt{3}/2$.

i, j, k

Complex numbers (see page 378) create an important link between the arithmetic of numbers and the geometry of the plane. Geometry is, in general, difficult for computers while arithmetic is straightforward. Questions about rotations, reflections, stretches and shears can all be translated into the language of numbers. In the nineteenth century, the Irish mathematician William Rowan Hamilton was trying to find a system of numbers that could do the same thing for three-dimensional geometry. He found that all the equations worked perfectly if he used four dimensions instead of three.

Hamilton's system of quaternions used three new numbers: i, j and k. Each of these numbers multiplied by themselves would give -1 (like the i of complex numbers) but they multiplied with each other in a very unusual way. For example, $i \times j = k$, but $j \times i = -k$. Quaternions remain important in computer graphics, spacecraft controls and robotics – all those fields which require efficient calculations of three-dimensional rotations.

Here as he walked by
on the 16th of October 1843
Sir William Rowan Hamilton
in a flash of genius discovered
the fundamental formula for
quaternion multiplication
$i^2 = j^2 = k^2 = ijk = -1$
& cut it on a stone of this bridge

William Rowan Hamilton was walking across Broom Bridge in Dublin, Ireland, when he had his revelation. He was so excited by his discovery of the equations for quaternion multiplication that he carved them into the bridge. Although the original carving has been eroded away, there is still a plaque on the bridge commemorating the event.

Infinity

In mathematics, infinity refers to something that is larger than any natural number, or a process that is increasing without bound. Infinity is not treated as a number, so it cannot be added, subtracted or multiplied in the way other numbers can.

The earliest written discussions of infinity come from ancient Greek philosophers. Zeno of Elea showed that Achilles could never overtake the tortoise because he would have to catch up to it an infinite number of times (see page 384). Aristotle argued that there were two kinds of infinity: the potentially infinite, such as numbers 1, 2, 3,… that never end, and the actually infinite, which has infinitely many elements but is complete, as in the number line between 0 and 1.

Believing in actual infinity can lead to some truly paradoxical results. Georg Cantor, in the twentieth century, showed that some infinities are larger than others (see page 404) and, indeed, that there is no biggest infinity.

The infinity symbol ∞ is called a lemniscate,
and was first used by John Wallis in 1655.

$\aleph_0$ (aleph-zero)

Aleph-zero is the smallest size of infinity. It is the cardinality (size) of the set of natural, or counting, numbers 1, 2, 3, 4, …. If a set has size aleph-zero it is called countably infinite. Sets of infinite size can exhibit some strange and counterintuitive properties. For example, the set of all even numbers is contained inside the set of all natural numbers, so you imagine it must be smaller. The natural numbers seem twice as big, since they contain both the even and odd numbers. Yet both sets have the same size: aleph-zero. You can see this by pairing each natural number n with the even number $2n$. For example, 1 is paired with 2, 2 with 4, 3 with 6, and so on. Since each element of one set has exactly one partner in the other set, and none are left over on either side, they must contain the same number of elements.

Many familiar sets have size aleph-zero, including the prime numbers, the integers (positive and negative whole numbers), the rational numbers (integer fractions) and the algebraic numbers (solutions to integer polynomial equations).

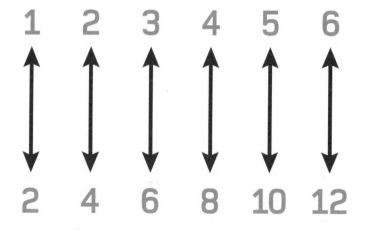

$\aleph_1$ (aleph-one)

The next smallest infinite cardinal after $\aleph_0$ (which describes the size of the set of integers) is named $\aleph_1$, or aleph-one. Georg Cantor proved that there were different sizes of infinity. His proof showed that the infinity of the real numbers (that is, decimals) is bigger than the infinity of the natural numbers, by assuming the two sets were the same size and then finding a contradiction. The set of real numbers has size $10^{\aleph_0}$, since each decimal has $\aleph_0$ decimal places, each of which has ten choices of digit. So we know that $10^{\aleph_0}$ is bigger than $\aleph_0$. But is $10^{\aleph_0}$ equal to $\aleph_1$? This question is called the continuum hypothesis.

In a proof that shocked mathematics, Kurt Gödel and Paul Cohen showed that the continuum hypothesis was independent of the axioms of mathematics. That is, it could never be proved nor disproved. Furthermore, we cannot even discuss the concept of a 'second-smallest size of infinity' without using the axiom of choice, another unprovable statement in mathematics.

Georg Cantor

ω (omega)

We use numbers in two ways: as cardinals and as ordinals. Cardinal numbers describe the sizes of collections; for example, I have three sheep, or twelve eggs. Ordinal numbers describe the order of a collection of objects; for example, this is the third sheep, or the twelfth egg.

In finite sets, ordinal numbers and cardinal numbers are the same, since we can count the objects in a set by putting them in an order. But with infinite sets the two concepts are separate. The smallest infinite ordinal is called omega, denoted ω, and it corresponds to the ordering of the natural numbers. Adding ordinals must be done carefully. For example, $1 + \omega$ is equal to ω, but is different from $\omega + 1$, which is a new ordinal. This is because starting with a number and then appending an ordered infinite set is different from first ordering an infinite set and then appending another number. Similarly, the numbers $\omega + 2$, $\omega + 3$ … are all distinct. Continuing the process gets us to $\omega + \omega$, and thence to ω^2, and eventually to ω^ω, which shows just how mind-bending numbers can get!

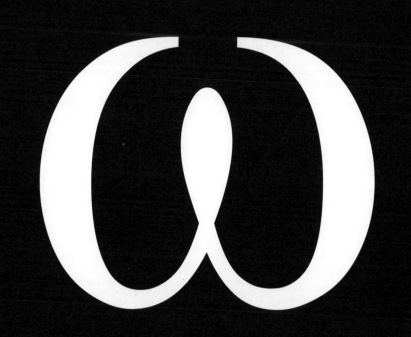

Glossary

Abundant number

A number that is smaller than the sum of its proper divisors. Twelve is abundant because its divisors (1, 2, 3, 4, 6) add up to 16.

Additive identity

A number that, when added to any other number n, gives the answer of n. In all our familiar number systems, the additive identity is 0.

Base

The number of digits needed to represent any number in a positional number system. For example, decimal is a base-10 system that uses the ten digits 0 to 9, while binary is a base-2 system that uses the two digits 0 and 1. The value of a digit in base b is determined by its position, with successive columns to the left multiplying the digit by powers of b. In base-10 the first column represents units (10^0), the second column the tens (10^1), the third column the hundreds (10^2), and so on.

Binary

A base-2 positional numeral system using two digits, 0 and 1, to represent numbers. Each successive column to the left multiplies a digit by two.

Bit

A binary digit. This means it takes two possible values, usually denoted by 0 and 1, or the on/off of a switch. Bits are the basic units of information in computers.

Byte

A byte is 8 bits, meaning it can encode 256, or 2^8, possible values. It is the smallest addressable unit of memory in a computer, and is sufficient to encode a single character of text.

Calculus

An area of mathematics that studies continuous change. One branch of it, differentiation, studies rates of change such as speeds and gradients, while the other, integration, studies curved areas and volumes. Both branches make use of the idea of infinitesimally small quantities and the notion of limits.

Cardinal number

A cardinal number describes the quantity of objects in a collection, for example, 1, 2, 3, 4, etc. (See, for comparison, Ordinal number.)

Complex number

A sum of real and imaginary numbers. Every complex number can be written as $a + bi$, where a and b are real numbers (on the standard number line) and i is the square root of −1.

Composite

A natural number that is the product of two smaller natural numbers. For example, 10 is composite because it is equal to 2 × 5.

Compound number

A number that is expressed using more than one unit. For example, in English, 'thirteen' is a compound number because it derives from the words 'three' and 'ten'. In French, the word for eighty, *quatre-vingts*, means 'four twenties'.

Constant

A fixed number that is unchanging. For example, in the expression $x^2 + ax + b$, the numbers a and b are constants, while the number x is a variable.

Cube number

Also known as a perfect cube, a whole number that is equal to three copies of a whole number multiplied together. For example, 8 is a cube number because it is equal to $2 \times 2 \times 2$, or 2^3 ('two cubed').

Decimal

The decimal number system is a base-10 positional numeral system. It uses ten digits and a decimal point. The value of a digit depends on its position within a number. (See also, Base.)

Digit

A numerical digit is a single symbol representing a number. For example, in our decimal system there are ten digits: 0, 1, 2, ... 9. Digits may be used in combination in a place-value system to represent bigger numbers; for example, 927.

Divisor

Also known as a factor, a number that divides into a natural number exactly, leaving no remainder. For example, 1, 2, 3 and 6 are divisors of 6.

Dual

In three-dimensional geometry, the dual of a polyhedron is the shape created by placing a vertex in the centre of each face and joining these new vertices with edges. In this way, the vertices of the dual shape correspond to the faces of the original, and the faces of the dual shape correspond to the vertices of the original.

Emirp

A prime number that is also a prime when its digits are reversed. For example, 157 is an emirp because both 157 and 751 are primes.

Emirpimes

A semiprime that is also a semiprime when its digits are reversed. For example, 26 is an emirpimes because both 26 ($= 2 \times 13$) and 62 ($= 2 \times 31$) are semiprimes.

Exponential factorial

The exponential factorial of a whole number n, denoted by $n\$$, is calculated by raising n to the power of $n-1$, which in turn is raised to the power of $n-2$, and so on down to 1. For example, 262144 is equal to 4$\$$, since it can be written as $4^{3^{2^1}}$.

Exponential notation

A number is written in exponential notation if it is expressed as $a \times 10^n$ for a real number a and an integer n. It is usually used for numbers that are very big or very small. For example, 43 million (43,000,000) is written in exponential notation as 4.3×10^7. Here the 7 indicates that the decimal place is to be moved to the right seven times.

Factorial

The factorial of a whole number n, denoted by $n!$, is the product of all the numbers from 1 up to n. For example, 4! is equal to $1 \times 2 \times 3 \times 4$, which is equal to 24.

Fermat prime

A prime number of the form $2^{2^n} + 1$. That is, we raise 2 to a power that is itself a power of two, and added one. For example, 5 is a Fermat prime because $5 = 2^2 + 1 = 2^{2^1} + 1 = 4 + 1$.

Fraction

A fraction is one number divided by another; for example, ½ or ¾. Fractions represent a whole divided into equal parts, so ¾ means three objects divided into four parts. The number above the line is called the numerator and the number below the line is called the denominator. (See also, rational number).

Hyperfactorial

A product of the numbers from 1^1 up to n^n for some whole number n. For example, 108 is a hyperfactorial because it is equal to $1^1 \times 2^2 \times 3^3 = 1 \times 4 \times 27$.

Imaginary number

A number that is the square root of a negative number. All imaginary numbers can be expressed as a real multiple of i, where i is the square root of -1. For example, $\sqrt{(-4)}$ is an imaginary number, and it can be written as $2i$.

Integer

Any number without a fractional part. In other words, the integers are the positive and negative whole numbers: $\ldots -3, -2, -1, 0, 1, 2, 3, \ldots$ and so on.

Interior angle

An angle between two adjacent sides of a polygon. For example, the interior angles of a square are each $90°$.

Irrational number

One that is not rational. This means it cannot be expressed as a fraction $^a/_b$ where a and b are integers. Examples of irrational numbers include π and $\sqrt{2}$.

Mersenne prime

A prime number that is one less than a power of two. Algebraically, Mersenne primes have the form $2^n - 1$ for some whole number n. For example, 7 is a Mersenne prime because $7 = 2^3 - 1 = 8 - 1$.

Multiplicative identity

A number that, when multiplied by any other number n, gives the answer of n. In all our familiar number systems, the multiplicative identity is 1 (one).

Natural number

The natural numbers are the positive whole numbers 1, 2, 3, ... excluding 0.

Ordinal number

An ordinal number describes the ordering of objects within a collection – that is, 1st, 2nd, 3rd, 4th, etc. (See, for comparison, Cardinal number.)

Palindromic number

One that reads the same forwards as backwards. For example, 12321 is palindromic.

Pandigital number

One that includes each of the digits 0, 1, 2, ... 9. Sometimes the zero may be omitted. For example, 8490271536 is pandigital.

Perfect cube

See cube number.

Perfect number

A number is called perfect if the sum of all the numbers that divide into it exactly (excluding itself) add up to itself. For example, 6 is perfect because its divisors are 1, 2 and 3, and $1 + 2 + 3 = 6$.

Perfect square

See Square number.

Polygon

A two-dimensional (i.e. flat) shape made of a finite number of straight edges and sharp corners. Triangles and squares are examples of polygons.

Polyhedron

A three-dimensional shape made up of flat polygonal faces, straight edges and sharp corners. For example, a cube is a polyhedron.

Power

A power of a number x means repeated multiplication of x with itself. So, seven to the power of two means 7×7, while five to the power of three means $5 \times 5 \times 5$. Taking powers, also known as exponentiation, is usually written using a superscript that denotes how many copies of x are being multiplied together. The examples above would be written as 7^2 and 5^3 respectively.

Prime

A natural number that is divisible by itself and one. For example, 7 is prime, but 6 is not because 6 is divisible by 1, 2, 3, and 6. Note that 1 is not prime.

Product

The product of two numbers a and b is the answer obtained by multiplying them together, $a \times b$.

Proper divisor

A proper divisor of a natural number n is a number that is less than n and divides exactly into n leaving no remainder. For example, 2 and 3 are both proper divisors of 12.

Ramsey theory

A branch of mathematics that studies order within large structures. A typical question in Ramsey theory asks how many people need to be at a party so that it is guaranteed that either three people know each other or three people are mutual strangers.

Rational number

A number is called rational if it can be expressed as a fraction of two integers, i.e. $^a/_b$ where a and b are integers (and b is not zero).

Real number

Any point on the number line. Real numbers are also referred to in this book as decimal numbers because they can be expressed as decimals (e.g. 84.10379), possibly with an infinite number of decimal digits. Whole numbers, integers and rational numbers are special kinds of real numbers.

Reciprocal

The reciprocal of a number n is the number $^1/_n$, or 1 divided by n.

Relatively prime

Two natural numbers are called relatively prime if there is no other number that can divide exactly into both of them. For example, 3 and 10 are relatively prime, but 4 and 6 are not because 2 divides exactly into both 4 and 6. (Note that two numbers can be relatively prime even if neither of them is prime.)

Repdigit

A natural number composed of the same digit repeated multiple times; for example, 11 or 999.

Root

A root of a number x is any number r where repeatedly multiplying r with itself eventually gives a value of x. For example, the square root of 9 is 3, because $3 \times 3 = 9$. The cube root of 8 is 2 because $2 \times 2 \times 2 = 8$.

Semiprime

A number that is the product of two prime numbers. For example, 15 is a semiprime because it is equal to 3×5.

Sexy prime

A prime that is separated from another prime by six. For example, 5 and 11 are sexy primes, because $5 + 6 = 11$.

Square number

Also called a perfect square, a whole number that is equal to the product of a whole number with itself. For example, 9 is a square number because it is equal to 3×3, or 3^2 (three squared).

Transcendental

A transcendental number is one that is not the solution of any polynomial equation involving whole numbers. That is, any equation created by adding, subtracting and multiplying a variable x and whole numbers. For example, π is transcendental, but $\sqrt{2}$ is not because it solves the equation $x^2 - 2 = 0$.

Twin prime

A twin prime is a prime that is separated from another prime by two. For example, 11 and 13 are twin primes, because $11 + 2 = 13$.

Vertex (pl. vertices)

In geometry, the vertex of a shape is a corner: a point where lines or edges meet. In graph theory, vertices are nodes between which edges are drawn.

Whole number

In this book, a whole number is used to mean one of the counting numbers 0, 1, 2, 3, . . . Note that in other texts, whole number can mean the same as integer, so includes both positive and negative numbers.

Index

Picture credits

First published in Great Britain in 2019 by
Quercus Editions Ltd
Carmelite House
50 Victoria Embankment
London EC4Y 0DZ

An Hachette UK company

Edited by Anna Southgate
Designed and illustrated by Dave Jones
Proofread by Rachel Mallig
Indexed by Patricia Hymans

A CIP catalogue record for this book is available from
the British Library

PB ISBN 9781787477315
EBOOK ISBN 9781787477308

10 9 8 7 6 5 4 3 2 1

Printed and bound in China